European Historiography
of Technology

ODENSE UNIVERSITY STUDIES IN HISTORY AND SOCIAL SCIENCES VOL. 156

European Historiography of Technology

Proceedings from the TISC-Conference
in Roskilde

Edited by
Dan Ch. Christensen

ODENSE UNIVERSITY PRESS 1993

European Historiography of Technology
is published with financial support from
The Danish Research Council for the Humanities

© The authors and Odense University Press 1993
Printed by Narayana Press, Gylling

Cover design by Ulla Poulsen Precht

ISBN 87 7492 910 0
ISSN 0078 3307

Contents

Acknowledgements 7
Editors Introduction 9
Akos Paulinyi: Machine Tools in the Transfer Policy of the Prussian 'Gewerbeförderung' (1820s-1840s) 17
John R. Harris: Sources for the Study of Industrial Espionage by Eighteenth Century France 30
Frank Allan Rasmussen: The Royal Danish-Norwegian Dockyard: Innovation, Espionage and Centre of Technology 41
Dick van Lente: The Critique of Industrial Technology in the Netherlands and other Western Countries in the Nineteenth Century 55
Helge Kragh: Transatlantic Technology Transfer: The Reception and early Use of the Telephone in USA and Europe 68
Hans Hedal: Early Rural Electrification in Denmark – a Reaction from People outside the Town Establishment 91
Klaus Mauersberger: The Development of German Engineering Education in the Nineteenth Century – a Comparison with Great Britain and France 111
G. Verbong: The Tension between Theory and Practice – Dutch Engineering Education in the Nineteenth Century 125
Michael F. Wagner: Danish Polytechnical Education between Handicraft and Science 146
Henrik Harnow: The Danish Engineer in Transition – The Reformation of Danish Engineering Education c. 1890-1933 164

Martyn Bakker: Farmers and Agrarian Industries in the 19th Century: Who's the Boss? 175

Dan Ch.Christensen: Danish Modernization Strategies – from above or below? 189

Peter Kemp: Ethics of Technology: The Irreplaceable 204

Kristine Bruland: Comparative Studies in European History of Technology 226

Contributors 235

Index 240

Acknowledgements

Publisher and editor gratefully acknowledge the generous financial support by the Danish Research Council of the Humanities in the preparation of the conference and its proceedings.

We also wish to thank all contributors for their cooperation in transforming conference papers to their present form. All contributions appear for the first time in the present volume.

I also owe my thanks to Odense University Press for their help in preparing the lay out and drawing the figures.

For any remaining faults I am all the more to blame.

D.C.C.

Editor's Introduction

Technology is generally understood as a transnational agent of fundamental importance for what we call the modernization process. But although technology is usually conceived of as an international phenomenon, a closer look on artefacts and regions will soon reveal fundamental differences between artefacts of the same catagory applied in various regions. Why were agricultural implements, steam engines, telephones, electrical networks and polytechnic institutions undergoing such intrinsic modifications? And why were the cultural structures of recipient nation-states affected so differently? These questions seem too complex to give a straightforward answer. Rapid technological change is part and parcel of the modernization process, but why are they, apparently – as history moves on – harmonizing formerly diversified cultures? If indeed they are? This is not so self-evident. Steam technology, for instance, would rather seem to heighten the contrasts between indigenous fields of production, originally generated by divergent national endowments.

It seems to be a natural thing for European historians of technology, so different in cultural background, to convene in order to come to grips with these questions by comparing their national processes of modernization.

Only recently has it been realized that the function of technology in the modernization process is tied up with national or regional political, socio-economic and cultural structures. As a consequence we are now facing national research teams sorting out the threads woven into their human fabrics. This research is not only trying to calculate the economic contribution of technology to the gross national product throughout history. The scope of research goes far beyond an attempt to open the black box of economic historians, who have more often than not explained technology's role in economic growth by merely mentioning it and taken its deterministic function for granted.

Economic historians have often been critized for treating technology's role in economic development too leniently when they claim that economic growth is determined by the application of cost-saving machinery paving the way for higher profits. Indeed this is a crude explanation and it is easy to repudiate it. It is equally plain to give examples of machinery motivated by cost-saving intentions that in fact turned out to be economically disastrous; and numerous examples indicate that technologies were introduced that were not even intended to improve the economy but enhance the prestige and power of royal courts and nation states. Still it remains a truism to assert that technology is the core element of the modernization process which is not only characterized by urbanization, rationalisation and secularization, but alas by unprecedented economic growth.

This criticism, perhaps, belongs to the past. Economic historians have long ago endeavoured to open the black box of technology. It is generally acknowledged that technology is a social or cultural construction, and the old assumption, that technological solutions are always available from the slot-machine whenever a capitalist desires to make new profits – market trends permitting – is obsolete. But opening the black box labelled 'technology' and finding 'a social construction' does not give much satisfaction. What does it really mean to say that technology is not only an artefact but knowledge and skill as well? And what is the implication of stating that a given technology is embedded in a given socio-economic structure and that consequently the transfer of a technology is provoking the cultural ambience of the recipient?

It does seem to imply, first of all, that technology is a historical phenomenon that must be studied in place and time. Once it has been recognized that the phenomenon 'technology' asserts an influence far beyond the range of an economic lever of riches, it becomes apparent that other aspects of technology's role in the modernization process have been given inadequate attention. It seems that to appreciate the role of technology in the modernization process it needs to be studied in its complex interrelationship with culture which is a category bound up with the history of nation states. To study the role of a certain technological development may very well end up with the highlighting of the peculiarity of a specific culture. And the study of the history of technology of various nation states will, hopefully, end up with the evidence required by historians, anthropologists and sociologists aiming at identifying patterns or at constructing theories of the modernization process.

During the centuries preceding the French Revolution the aspirations to develop technology were often concentrated on the perfection of military and naval technologies to favour the international position of nation states in Europe. Mercantilist policies of technology aimed at securing advantages of one nation state to the detriment of rival powers. Colonial expansion, royal prestige and a favourable balance of trade were political goals in need of technological competitiveness.

Agricultural improvers complained that so much effort had been invested in status and warfare to the detriment of ample food production for increasing populations. Outside Britain, governments enticed foreign technicians to offer their services to their kingdoms in return for exorbitant premiums thus depriving their rivals of their technical skills. Mercantilist policies have often been ridiculed by economic historians in favour of a liberal capitalist doctrine, but goals and resources of the royal court differed from that of the private capitalist in an open society. Moreover, their enterprises were conducive to technological development at a time when private capital accumulation was insufficient and a surplus within a time span of a hundred years was a promising perspective in the eye of the contemporary observer. Technological dirigisme so dominant in the mercantilist predatory countries was completely unthinkable in the heartland of the industrial revolution, where the greatest source of encouragement was the personal interest of its inhabitants. Only later on did European nation states attempt to emulate Britain by adopting her liberal policies, but the governments of the late-coming nations continued to give a helping hand in closing the technological gap by organising the transfer of machinery, the emigration of mechanics, the industrial espionage, and the education of technicians, etc. This is to illustrate firstly that technology has to be studied historically, i.e., relative to time and space, and secondly that the history of technology amounts to international communication by historians who are digging up the past of technicians who in their day communicated internationally.

Let me indicate by just one example the implication of the statement that technology is a cultural construction. The delicacy of this uncontroversial assertion is that technology is a human creation and consequently embedded in a certain culture. Of course technology cannot be separated from its cultural embeddedness because technology is not only an artefact but also the knowledge and skill and purpose involved. The swingplough consisting of a frame and an iron cast mouldboard was a British construction made possible by the coke smelting process and originating in a market dominated by large capitalist farmers employing wage labour. British capitalist agriculture was having the leading edge at the time in Europe. The swingplough was a costly production implement, but advantageous because it was more productive and less energy demanding than the traditional tool. When transferred to Denmark the swingplough was operating in a society with a dual agricultural production system, viz. an estate-economy aspiring to emulate the British post-enclosure-system and a peasant economy still producing mainly for selfsufficiency. The British swingplough was given the function of a symbolic vehicle to establish similar socio-economic conditions thus wiping out the existing peasant economy. However, the Danish peasantry proved to be sufficiently strong to shape the transferred technology to serve their own ends.

The outcome was not only a modification of the swingplough. It was also a successive change of the agricultural system of the recipient country in a way which nobody was capable of predicting. The cultural construction of technology does not imply that either the technology challenges the socio-economic structure or vice versa. The transfer of technology will rather affect both. And this is what makes empirical studies of technology transfer so intriguing. All along the controversy about the modern British swingplough took place as if it were merely a question of a technological artefact, but behind this a great many farmers were well aware that the artefact played the symbol part of a strategic conflict of interests. A separation of an artefact from its cultural ambience seems an impossibility. Consequently it is an oversimplification to look for the cultural effects of a given technology, because the technology is essentially a cultural product right from the outset, although maybe from a different culture. There is no clean culture-free technology. And there is no modern culture untouched by massive technological impact.

The present volume contains a series of contributions to illuminate the vast field of 'the cultural construction of technology', technology transfer and polytechnical education. The papers were given at a conference in Roskilde in February 1992 attended by some 40 historians of technology from Denmark, England, Germany, the Netherlands, Norway and Sweden. The participants at the conference are all engaged in studies of the modernisation processes of nation states and the purpose of the conference was to discuss the transfer of technology in a comparative European perspective.

In recent years national histories of technology have been or are being written in Europe. In 1989 a Swedish history of technology[1] was published and the same year Joachim Radkau discussed the peculiarity ('Sonderweg') of 'Technology in Germany from the 18th century to the present time'.[2] In Holland a group of historians of technology is about to finish a history of technology in the Netherlands in the 19th century in four volumes to be continued by another work covering the present century.[3]

The Norwegian Research Group, the TMV-Centre on technology and human values is deliberating a similar project, although TMV-Sentret is devoted to interdisciplinary studies of the relationship between human values and technological development including a wide range of topics such as lifestyles, technological risks and environment studies, ethical issues of scientific research and the engineering profession, topics being usually connected, but not necessarily restricted to technology philosophy, because these fields, of course, are subjected to historical change.[4]

Finally the Danish TISC-project (Technology, Innovation and Society in a Cultural Perspective), the organizer of the conference, is engaged in writing a

Danish history of technology and culture in 3 volumes covering the period 1750-1990 to be published in 1994.[5]

There may be several other works in progress that I am not aware of, but those mentioned testify to the fact that national histories of technology are mushrooming. Of course the historiography of small nation states is a more practicable endeavour than writing a comprehensive history of technology of say the UK or the USA. A British colleague of mine aired the view that a comprehensive history of technology was only imaginable for small nation states where technological innovations are scarce and the cultural and socio-economic ambience is so relatively confined as to allow in-depth investigations. Although I saw his point I am not so sure that the scarcity and clarity compensates for the complex task of investigating archival material from at least two nations (one being often the UK) to study technology transfer.

However, some preliminary results by some individual historians of technology as well as by members of some of the above project groups were discussed and sometimes even compared at the conference. Here are some indications of what the various articles contain.

A discussion about the concept of technology transfer marks the introduction of Akos Paulinyi's paper. One of the basic merits of his study, it seems to me, is his persistent quest to come to grips with the question, 'how did they actually make it?'. This is by no means a return to old-fashioned bolts-and-nuts-history of technology. By contrast it is a method to understand how technology transfer was in reality performed. The outcome is a fascinating history of international communication between inventing and predator countries comprising all aspects of protection against technological espionage, transfer of machine tools and diffusion by means of public workshops and education of mechanics. John R. Harris generously shared with his younger colleagues his intensive experience from working with French and British source material collected by public officials serving opposing interests of producers and spies of technological innovations. This material, far more abundant than commonly realised, reflects government initiatives and assessments of technological priorities. Like Paulinyi, John Harris stressed the observation that technology never passed by writing from country to another, but eye and practice alone were capable of training men in these activities. A series of examples of industrial espionage by the Danish Royal Naval Dockyards during the 18th century was delivered by Frank Allan Rasmussen. He makes it increasingly obvious that the noble art of industrial espionage was a comprehensive key activity in early technology transfer, the methodology of which underwent considerable refinement.

The problem of cultural restraints in the acceptance of technologies was, in different perspectives, the subject of papers given by Dick van Lente, Helge

Kragh and Hans Hedal. Van Lente considered the limited Dutch criticism of, and resistance to, the new technologies imported during the 19th century, and argued that the Romantic-Conservative basis of this criticism was a major reason for its failure to influence the country's technology policy. Dick van Lente's plea for the relevance of ideology in understanding the introduction and integration of new technologies in society was restated by Hans Hedal, who considered the Danish physicist Poul la Cour's attempt around 1900 to develop a decentralized electrical system based on wind power and the factors determining its diffusion process within the country. Hedal demonstrates how the introduction of urban electricity networks threatens to disfavour the interests of the newly established cooperative dairy movement, responding by associating around the development of an alternative electrical system appropriate to serve their ideological and economic strategy. Whereas van Lente and Hedal dicussed cases of inter- and intra-national transfer and diffusion, Helge Kragh discussed the different developments of early telephone technology in the USA and Europe. Arguing that the slow European response to the telephone reflected cultural values particular to the Old World, he suggested a general model for understanding how new technologies come into existence by gradual fixation of the socio-cultural space.

In the session on engineering history, Klaus Mauersberger discussed German engineering education in the 19th century in comparison with French and British. Mauersberger argues that conflicts about contents of curricula, most notably, perhaps, the question about theory and practice, not only reflect divergent cognitive issues, but more wide-ranging societal and professional interests. This conclusion seems to cover engineering institutions in the Netherlands and Denmark as well. Geert Verbong traced the origins of Dutch civil engineering education from specific national demands for waterways, land reclamation and protection, arguing that the Dutch case fits the French-German model characterized by academic training of civil servants for the state bureaucracy. Michael F. Wagner and Henrik Harnow both dealt with the history of Danish polytechnic education from 1829 to the 1930s. Particularly when compared with Prussian technological policies (cf. Paulinyi's paper), it becomes obvious how damaging the onesided stress on science and the correspondingly low priority given to workshop training must have been to Danish industrial development. Why was this so? Wagner concludes on the basis of a close scrutiny of the source material related to the first Danish Technical University (PL), that its foundation was motivated by ambitions to strengthen the position of the would-be science faculty at the University of Copenhagen rather than to educate practising mechanics and engineers for private enterprises. Henrik Harnow traces the successive changes of the PL-curriculum and concludes that by the turn of the century the high priority given to science subjects eventually turned out advantageously due to the increased emphasis on theory related to the modern branches of civil engineering, i.e. electricity networks and concrete constructions.

Technology transfer and innovations in agriculture, traditionally key subjects in both Dutch and Danish histories of technology, were treated by Martyn Bakker and Dan Ch. Christensen. Bakker stressed the importance of group-culture and mentality in his paper on innovations related to sugar beet refining and cream separators in Dutch agriculture. He concluded that rationality for big capitalist manufacturers is quite a different concept from that of small peasant producers whose response to technology innovation can only be understood by taking into consideration their social and mental attitudes so closely tied up with opposed socio-economic interests. Christensen argued along similar lines in his case stories about the transfer of a new cultivation system and swing ploughs from the UK into Denmark around 1800. The receptivity of these innovations depended heavily upon socio-economic interests and traditions of technology control. As a third case he took up the success story of the Danish cooperative movement trying to explain its momentum by analysing its technological components within the same theoretical framework.

In addition to these more specialized papers, a somewhat broader discussion on the ethical implications of living in a technology dominated world was introduced by theologian and philosopher Peter Kemp. Finally Kristine Bruland concluded the conference by presenting a richly textured address on the problems and promises of doing comparative studies of European histories of technology.

An important outcome of the presentation and discussion of these papers, and I think that the participants of the conference would agree to this, was the increasing awareness that because technology transfer was and still is a matter of international communication, the study of this is in need of networks linking the different cultures. We shall meet again as the current studies of national histories of technology have reached a stage enabling us to make in-depth comparisons. This, hopefully, will contribute strongly to a more penetrating understanding of the European process of modernization.

Dan Ch. ChristensenRoskilde, August 1992

Notes:

1. Jan Hult, Svante Lindqvist, Wilhelm Odelberg, Sven Rydberg, 'Svensk teknikhistoria', Gidlunds Bokförlag, Hedemora 1989, reviewed among others by Francis Sejersted in 'Polhem' 8, 1990.
2. Joachim Radkau, 'Technik in Deutschland. Vom 18. Jahrhundert bis zur Gegenwart', Neue Historische Bibliothek, edition suhrkamp, Frankfurt am Main, 1989.
3. See Dick van Lente, 'Work in Progress: Technology in Dutch Society during the Nineteenth Century', in Tractrix 2 – Yearbook for the History of Science, Medicine, Technology and Mathematics, Utrecht, 1990.
   'Geschiednis van de Techniek in Nederland', vol. 1-6, Walburg Pers, Zytphen.
4. Further information available from
   TMV – Centre of Technology and Human Values
   Forskningsparken, University of Oslo
   Gaustadalléen 21
   N-0371 OSLO 3
5. Further information available from
   TISC-Project
   University of Roskilde
   House 3.1.5.
   P.O. Box 260
   DK-4000 Roskilde

Machine Tools in the Transfer Policy of the Prussian 'Gewerbeförderung' (1820s-1840s)

by Akos Paulinyi

My paper is focused on some general remarks on the problem of technology transfer and a case study from the broad field of governmental support in technology transfer in Prussia. The general remarks are concerned with the role of printed information and of personal contacts in transfer and with the impact of the British statutes prohibiting the exportation of machinery. The object of the case study is the import of machine tools from Britain and the support to produce them in Prussia.[1]

Let me start with explaining what I mean by speaking of technology (in German Technik) and of its transfer. The most general intention of technology, the most general objective of every technical action is the change of the state of material, energy and information. The core of technology is an artificial object (a simple tool or a complicated technical system for processing materials produced by man); nevertheless, technology means not only a heap of artefacts but a system of artefacts as well as processes and actions, in which man designs and produces these artefacts as well as uses them to achieve a certain end. There is no technology without man and hence, because man is a social creature, technology cannot be independent from society and its socio-economic system.

By technology transfer is meant the acquisition and transportation of a technology from one place where it already exists to another place where it does not yet exist. The recipient place is separated from the place of origin by state borders. The final purpose of the transfer is to implant the acquired technology into the existing system of technology. But technology cannot be independent from society, and hence technology transfer means an implantation into the existing economic, social and cultural system. The crossing of state borders is one important feature of the transfer of technology; another one is the initiative, the activity of the recipient country in order to acquire and implant the new technology. In economic terms transfer always means (legal or illegal) imports. This is not to say that every export of technology from a country of origin is equivalent to technology transfer, even if every export of a new technology also implies its diffusion. I would prefer to distinguish between technology transfer and transfer policy on the one hand, and between exports and export policies on the other in order to penetrate new markets and to overcome trade barriers.

To sum up: Technology transfer is a boundary-crossing move and comprises in my understanding all the activities undertaken by the recipient country. A policy of technology transfer includes all means and methods of acquiring and implanting the new technology.

It is a truism, that technology transfer – as Timo Myllyntaus puts it – "is not just a matter of moving some piece of hardware from one place to another"[2] and every transfer is certainly "a complex process of transformation that necessarily takes place in a specific economic, social and cultural context" (whatever that means). But in view of the fact, that it became very fashionable to stress the context and forget the hardware, I would like to point out, that the "moving of some piece of hardware", the acquisition of a new device, machine or technique of production was and is the core of every technology transfer.

By diffusion of technology we understand the spreading of an already existing technology inside the boundaries of a country. It makes no difference whether the already existing technology was invented in or transferred into the country in question. To put it in another way: Diffusion as I use this term means the spreading of a new technology in the state territory, where it was previously invented or transferred.

The only reason for this extensive explanation of the basic terms frequently occurring in this paper is to avoid misunderstandings. It is not my intention to start an argument on terminology. I am aware, that in Anglo-American research the term diffusion is used – as international diffusion – for what I call transfer and as internal diffusion for the spreading within a country. On the other hand: in recent German publications on innovations it is very fashionable to speak about technology transfer even when a new technology is moved from Stuttgart to Hamburg.

Technology transfer is not an event, but a process. It is not a self-acting process and this is one of the reasons why I wish to avoid the term diffusion which makes the impression of something self-acting. Apart from ressources available, the success of transfer is mostly dependent upon the insight in the problems to be solved, upon the choice of methods and upon the intensity of activities. Transfer is initiated by getting information about the new technology, followed by the most distinctive feature of the process, that of acquiring the new technology and ending up with the innovation, i.e. the durable, long-lasting use of the acquired technology within the socio-economic system of the recipient country. To achieve this stage is not possible without spreading the new technology and hence in all transfer policies we know, the support of transfer and the support of diffusion were closely linked.

I am fully aware of the impact of existing economic, social and cultural structures on the success or failure of technology transfer. Nevertheless, the focus of my paper is "on the moving of some piece of hardware" in a very narrow section of technology transfer in Prussia between the 1820s and 1840s: on the methods of governmental activities and the role they played in order

a. to acquire machine tools in Britain and to implant them into the existing technology of metal working, and
b. to start the production of machine tools in Prussia.

Before we enter this narrow field, let me point out a few general (non specific Prussian) problems of transfer. This seems to be useful, because despite the differences of the social, economic and cultural structures of the recipient countries, including the difference in the level of technical skills and knowledge, despite the variety of means and methods for acquiring the new technology, there were a few very significant and common features of transfer during the first half of the 19th century.

Firstly: after 1815 even the most advanced continental countries like France, Belgium, Switzerland, Prussia and Saxonia had to close a technological gap much wider than before 1790.

Secondly, up to the 1840s technology transfer was almost a one-way traffic from Britain to the continent. The very few exceptions are only proving this generale rule.

Thirdly, transfer, as Cameron puts it "was highly dependent upon personal knowledge and contacts".[3] I would like to add: Dependence upon personal contacts not only between persons, but dependence upon the personal contact with the technology in question; and dependence not only upon knowledge, but probably more important, upon the skill of the people involved. This high dependence upon personal contact with the new devices, machines etc. was inevitably due to the "nature", to the novelty of the technology developed in Britain on the one hand and the traditional, purely empirical system of technical training in the crafts on the other hand.

Finally: up to 1843 all countries interested in British technology had to overcome the trade barriers based on the statutory ban against the exportation of machinery and, up to 1824, also on the act illegalizing the emigration of artisans.

Concerning the importance of personal contacts, the role of printed information and the impact of the statutes mentioned above let me make a few remarks:

Even now transfer of technology is still dependent upon personal contact with the new implements and upon the participation in the process of production. This dependence was much stronger during the first half of the 19th century, as the overwhelming bulk of British technology was not intelligible through written (or printed) information or through drawings published in Britain. Hence it was very difficult to communicate technical knowledge or acquire technical skills outside the production process itself (i.e.: outside the technical action) . This seems to be a consequence of the way new technology came into being.

As we know, the bulk of new technology was of "non scientific origin" and created by practicians. The new technology was created in Britain with the old hand-tool-technology, but by thinking about new solutions to a problem and not by thinking about improving old solutions. The new machine-technology was created by highly skilled and gifted artisans. From the traditional point of view they were bad artisans: instead of using the old and therefore well working methods in executing a technical task they were thinking about new ways, thinking away from old well known methods and looking for something new.

In accordance with the traditional methods of technical education of skilled artisans in the practice of production (learning by doing) the only way to conceive and understand new technical solutions (or new combinations of old solutions) was the personal contact with new technology. This is not to say, that the level of technical skills of artisans in France or Germany was lower than in Britain about the 1760s and 1770s. It has very much to do with the principles and traditions of crafts. Everywhere craftsmen were educated on the job; training in manufacturing technical products from drawings and descriptions was unusual. The world of craftsmen was mostly a world of real things they could see and touch. Drawings were based on real existing things or on own ideas. There is a big difference between producing a drawing or a sketch on the basis of one's own ideas and producing a new device, one has never seen, on the basis of drawings or descriptions. Therefore even for the highly skilled, experienced craftsman, although well trained in making and reading drawings, the personal contact with the artifact, the "kindergarten method" (show and let me see), was the best way to understand new technical solutions and to learn to reproduce them.

As to the role of printed information there was indeed since 1820 a flood of periodicals spreading news and even information on technology and science. On the basis of the density and frequency of printed information and of the rapidly growing variety and quantity of technical journals on the continent Hans Redlich concluded, that the role of personal contacts in technology transfer was diminishing since the 1830s.[4]

This allegation, followed by similar statements from other German historians, is overlooking the origin of printed information and in face of their growing quantity overestimating their quality. Firstly; only very little of the printed information was a substitute for personal contacts; to the contrary, most of it was the result of such contacts. The bulk of the marvellous drawings and of printed documentation about British technology, published in journals on the European continent, was the product of an already realized transfer. Most of the well documented articles were not incentives for but messages about the transfer. Nevertheless; the growing number of well documented information was a very important contribution to the spreading of knowledge on new technology. Secondly, the value of printed information, patent specifications included, cannot be proved by counting their frequency. It can be checked only by reading them. Thirdly, information published, say 1830, and very valuable to the historian of technology hunting the last missing link in the chain of his arguments, would have been useless for a contemporary craftsman interested in transfer. One proof only:

Searching the paths of the transfer of a hot-blast from Scotland (in 1828/29) to Baden-Württemberg (in 1832/33) I collected 18 articles published between January 1829 and April 1833 in 8 British, French, German and American journals. The first and second publication contains a very short and misleading piece of information on the patent Neilson took out 1828/29. Even in 1831 7 articles are repeating from different sources the information on the hot-blast experiments in Clyde from 1829. In the articles 9-18 there is new information about the output and the cost-reduction achieved by the hot-blast stove. But up to 1833 there was no information on the design or on the construction of the heating apparatus and the publication only gave the first hint that something happened. Typically, the first details and drawings from the construction of the heating apparatus in Clyde was based on personal contact and published in France in the last months of 1833.[5]:

Printed reports from the hot blast in Scotland
(from January 1829 to April 1833)

No	year/month	published in	reprinted from
1	1829/01	DPJ 31/1829	?
2	1829/08	DPJ 33/1829	Register of Arts 1829
3	1830	MM No 334/1830	Glasgow Chronicle
4	1830/02	DPJ 35/1830	MM NO 334/1830
5	1830/03	BST XIV/1830	Le Temps 25.3.1830
6	1830/06	BST XV/1830	Revue de revues
7	1830/10-12	JTCH 9/1830	BST XV/1830
8	1830/10	BST XVI/1830	BISA 6/1830
9	1830/11	MM No 379/1830	Boston MM
10	1831/01	DPJ 39/1831	MM No 379/1830

(Continued)

No	year/month	published in	reprinted from
11	1831/04	DPJ 40/1831	BST XVI/1830
12	1832/04	EHSc VI/1832	
13	1832/06	MM No 460/1832	
14	1832/08	DPJ 45/1832	MM No 460/1832
15	1832/08	DPJ 45/1832	EJSc VI/1832
16	1832/12	DPJ 46/1832	MM No 467/1832
17	1833/02	RPI Februar 1833	
18	1833/04	DPJ 48/1833	RPI Februar 1833

List of journals:

Boston MM	Boston Mechanical Magazine (USA)
BISA	Bulletin industrielle de la Société d'agriculture de St.Etienne
BST	Bulletin des sciences technologique. Ve Section du bulletin universelle
DPJ	(Dinglers) Polytechnisches Journal
EJSc	The Edinburgh Journal of Science
JTCH	Journal für technische und ökonomische Chemie
MM	The Mechanics' Magazine, Museum, Register, Journal and Gazette (London)
RPI	Repertory of Patent Inventions

Concerning the efficiency of barriers built by British authorities against the outflow of machinery and artisans and backed by the lobby of the textile industry we should differentiate in time and in products and should not confuse the existence of a law with the working of this law. Starting with the wars against France from the 1790s up to 1814/15 the violation of current laws could be interpreted as high treason – as espionage in the narrow sense – and the emigration and/or exportation of machinery was indeed a risky undertaking. But after 1815 smugglery of men and machines and spying (collecting information) on factories in Britain had become less risky and the barrier to be overcome in "espionage" was the owner rather than the prevailing law.

After 1824/25 when the law against the emigration of artisans was abandonned and a new law on the exportation of machinery introduced a complicated licensing system, both the acquirer and supplier of whatever machinery had the game against the Board of Trade and the Customs Offices in their own hands. It was not only the geography of the British Isles – making a perfect control of shipping impossible – that worked to the advantage of smugglers. Exporters had their support in the Board of Trade, being "a haven for freetraders"[6] and were unconsciously backed by the lack of staff and technical knowledge of the customs service. The enforcement of the regulations, as S. Pollard puts it, "exceeded the administrative capacity of the day"[7] and I would add: it exceeded far more the "technological capacity", i.e. the knowledge about technology of the administrators. This is not to say, that customs officers were ignorant or stupid. But they were faced with a rapidly and constantly growing number of consignments, of a variety of artefacts on the one hand and confusing and outdated articles of the regulations on the other.

Exporters tried one of their many tricks to fool the officers: they mixed parts of allowed and of banned machinery in the same consignment, they changed the numbers of packages, in their applications to the Board of Trade they used misleading names of the parts and described them in words not mentioned in the statutes etc. Consequently Customs officers had to call for the expertise of a machine-maker playing the part of an independent expert in cases where his fellow machine-makers were involved. And if the illegal nature of the consignment was too selfevident making the friendly expert unable to turn a blind eye, it happened very often that in the following investigation the officers of the Board of Trade decided in favour of the exporter.

To demonstrate the reluctance to prosecute every illegal export let me tell an unbelievable story from the records of the Board of Trade.[8] The British consul in Göteborg, Mr. H. Th. Liddell reported to the Board of Trade that the British coaster "Nancy" had arrived on 25th February 1835 in Göteborg with 43 cases of "valuable textile machinery for spinning yarn". At the first interrogation the captain Mr. Marflet related, that in Newcastle he had met a Mr. Hennig from Göteborg and the consignment had been loaded under the supervision of custom officers. In April the Customs Office in Hull confirmed, that the consignment had been loaded in Hull, shipped to Newcastle and exempted from the customs regulation; now the captain did not remember anything, he had never heard from a Mr. Hennig and had never spoken to the consul in Göteborg. He stated, that he "sailed for Newcastle on 14th February last, but was forced by distress of weather into Gothenburg". On 9 May the Board of Trade concluded: "Mr. Liddell's statement is too peculiar and clear to admit of Marflet's contradiction. The question is whether a prosecution can be successfully carried on – and this is a matter for the Treasury and Customs".

In the light of the research of Chaloner[9] and Jeremy this example of reluctance is not casual. In my opinion the role and impact of the laws prohibiting and, since 1825, regulating the exportation of machinery has been exaggerated. This seems to be true in general and in particular for the time after 1815 and 1825 resp. The overestimation fits very well into the picture of the heroic and risky role played by men involved in technology transfer. Up to 1842 the existing laws were a nuisance, both for suppliers and buyers of machinery[10], but after 1815 they yielded diminishing returns and were no serious barrier to the transfer.

Let me come to my short case study starting with a brief outline of the Prussian institutions of transfer.

In the course of reforming the Prussian state administration after 1808 the Department of Trade and Commerce (in German: Abteilung für Handel- und Gewerbe) changed several times from the ministry of internal affairs to the ministry of finances (in British terms from the Home office to the Treasury). After its foundation in 1808 the Gewerbe-Department disposed of yearly 65.000 Taler "for the encouragement of trades by better equipment and better tools", and since 1828 of 100.000 Taler (after 1873 = 300 000 Marks = £ 15 000).

The head of the Gewerbe-Department from 1818 to 1845 was Peter Christian Wilhelm Beuth (1781 – 1853). In order to support the Gewerbe-Department's technological decisions the "Technische Gewerbe und Handels Deputation" was founded in 1811, an advisory board, consisting of "civil servants, scholars, artists, farmers, manufacturers and tradesmen". The Deputation was to study abroad the development of "Gewerbekunde", i.e. of the knowledge of commerce and industry and had to make public use of this knowledge. It was reorganized under the impact of Beuth's criticism in 1818 and next year Beuth, head of the Gewerbe-Department, took over also the post of the director of the "Königliche Technische Deputation für Handel und Gewerbe".

In 1821 Beuth completed his system of "Gewerbeförderung" by founding the "Verein zur Förderung des Gewerbefleißes in Preussen" (Society for the Encouragement of Industry and Trade) and by establising in Berlin the "Technische Schule" (since 1827 "Königliches Gewerbeinstitut") for the training of technicians. From the 1820s Beuth was head of all governmental and state-supported institutions involved in the transfer and diffusion of technology (1829 he also took over the directorship of the "Bauakademie, founded in 1799), constituting one of the most important activities of "Gewerbeförderung".[11] The activities of the "Verein", publishing from 1822 the "Abhandlungen des Vereins zur Förderung des Gewerbefleißes in Preussen" were focused on the diffusion of technical knowledge. The "Technische Schule" was geared towards the production of technological education; it started as a "better school" for craftsmen and developed in the late 1820s into a school for technicians of textiles and metalworking.[12] This type of technical schools, founded in the 1820s and 1830s in most of the German states, concentrated on the theoretically based but mainly practical education of skilled mechanics. It played a very important role in closing the technological gap between Britain and Germany. To achieve this goal in the 1830s and 1840s the practical training at the workshops of the schools with modern technical equipment was more important than the scientific-theoretical approach. This scientific approach was a long-term investment: apart from chemistry it yielded returns only after the 1860s, when different branches of technology gradually became more and more science-based (and the technical schools turned into technological universities).

In order to close the technology gap between Britain and Prussia/Saxony, the main attention was paid to transfer of textile and power technology, and partly to flour-milling and from the 1820s to railways. The first precondition for starting any activity was information on the existence of the new technology and in order to start the transfer collection of more specific technical information on the nature of the new technology was required. Collecting information by reading foreign "technical" journals, patent registers and printed reports from "tours" was the main business of the "Deputation" and very important for preparing visits in Britain. The bulk of information in technological journals, e.g. the Prussian Verhandlungen, different Saxonian and Bavarian journals and transactions, the French Bulletins as well as the most important private journal, Dinglers Polytechnisches Journal, was dealing with textiles, steam engines, mechanical engineering, railways, and most of the money spent by the civil service on transfer went into those fields.

The first aim, the acquisition of new technology, could be achieved by buying the objects (machines, devices and even materials, e.g. coal, pig iron for foundries, machine yarn for power-loom weaving), by hiring skillful and knowledgeable persons, and by getting them out of Britain. In the beginning of textile transfer it was almost inevitable to buy both the machines and the men, who were able to operate and maintain them and to train native workers and artisans both in the running of machines and in the repair of them.

Almost at the same time as the implantation of new technology took place, the most advanced countries on the continent started to tackle the second task, i.e. the production of the new technology. This proved to be a decisive step in order to reduce or eventually to abolish the dependence on foreign suppliers and it appeared to be a longer lasting venture than the import of machinery. The difficulties involved were many, beginning with acquiring the basic information on machine tools. Even to well informed and educated outsiders, i.e. people not working in the trade, it was not an easy task to recognize the importance of machine tools. Up to the 1820s there is a lack of printed information in Britain; the first builders of machine tools (the generation of Maudslay, Roberts and Fox) did not take any patents, none of them published anything on the machine tools they produced.

The first mention of a new lathe made by Mr. Maudslay was published by Gregory Olynthus in 1806, a short description of a lathe of Mr. James Fox in 1813.[13] Therefore it is really fascinating to see, that on his first journey to Britain in 1823 the well informed Beuth, being very sensitive to the impact of machine tools for creating a machine building industry in Prussia, had the most outstanding producers on his itinerary visiting among other manufacturers also Fox, Maudslay, Holtzappfel and Murray.[14]

In the transfer of machine tools from Britain to Prussia and their diffusion the paths and methods were the same as in other branches:

1. Once the information was given, Beuth (i.e. the government) bought the machine tools in Britain, to use them in the workshops of the Gewerbe-Institut or to lend them to private manufacturers.
2. The state supported journeys of civil servants or of students of the Gewerbe-Institut mostly to Britain in order to collect information or to be trained "on the job" in well equipped establishments.
3. The state eased the spreading of specific knowledge and of more general information about the machine tools by subsidizing the publications of detailed descriptions and drawings preferably in the "Verhandlungen".
4. By lending modern machine tools to private manufacturers to improve their equipment or by subsidizing journeys the state supported the foundation of engineering firms.

The acquisition of the machine tools from Britain for the workshop of the Gewerbe Institut in Berlin was the most important step both for the transfer and the diffusion of modern technology in metal working. What counted in order to close the technological gap was not the quantity but the quality of the acquired machine tools. In 1826 the Gewerbe-Institut owned three "English lathes". Between 1831 and 1833 the Gewerbe-Institut acquired a large self-acting lathe, two lathes for facing and a planing machine from Fox in Derby. From Roberts in Manchester they bought a wheel-cutting-lathe in 1835 and a drilling and a screw cutting machine in 1843. Even if the diameter of their screw exceeded 1.5 inches and therefore, according to the statute 89 ex 1786, their exportation was prohibited, there was no trouble about the export from Britain; all the machines bought by Beuth were ordered by His Majesty the King of Prussia from His or Her Majesty the King or Queen of England for teaching purposes, i.e. for exhibition in the Gewerbe-Institut in Berlin.

The imported machine tools served in the mechanical workshop not only as exhibits but for the practical training of the students in metal working. To have the real machine at disposal was a precondition for the next step, its copying. In order to learn the principles of construction and the manufacture of machine tools the acquired machines were completely dissassembled. From the parts were made copies, produced drawings and models. The technical documentation, the detailed descriptions and excellent drawings produced from the machines and their dissassambled parts were published in the "Abhandlungen". The public had free access to the documentation, to the models, built by the students and exhibited in the model collection of the Gewerbe-Institut and both the imported machines and the models could be copied by anybody interested in metal working. The open access to all visitors was one of the basic conditions also for lending modern machinery to private manufacturers. The machine tools, the documentation and the models, the training of mechanics on the best machine tools in the Gewerbe-Institut and the machine tools lent to workshops contributed to the spreading of knowledge in metal working.

It is very difficult to measure the impact of this government support on the speed of technology transfer and of the diffusion of machine tools. In the 1840s the Prussian administration spent only about 40.000 Taler (120.000 Mark = £ 6000) on the import of machine tools. According to price lists for machine tools 40.000 Taler were the price for 20 larger or 40 smaller "English lathes"(i.e. iron built lathes with a slide rest). Compared with the price of a steam engine or with the price of implements even for a small spinning factory the money spent on machine tools seems to be a quantité negligeable. But to measure their importance in money terms and comparing the lending of one machine tool with that of one mule is conducive to the point: to start a spinning factory in the 1830s one would want a set of machines for carding, roving and spinning. Metal working was different; even one "English lathe" or whatever modern machine tool could help a skilled craftsmen to start the production of other machine tools. The presence of even one machine tool was much more important in closing a technological gap than financial support.

Whether the gap was eventually only less wide, narrower or even closed we are not able to measure. But the big leap of the German machine building industry in the 1840s and 1850s can be proved by the output of the most outstanding product of machine builders: the locomotives.

It is well known that during the first years of railway construction the machine builders in different German states were not able to supply locomotives. But after 15 years the locomotive builders in Berlin, Chemnitz, Hannover, Kassel, Karlsruhe and Munich dominated the domestic market, started to export locomotives and to compete with British and American products on the European market.[15] To achieve this goal in competition with the British machine builders was possible only by means of modern machine tools and of workmen and technicians they could handle them. This is not to say, that this big leap was caused by government support of the transfer of machine tools only. But we cannot rule out the impact of the government transfer policy including not only the support of imports but also the training of mechanics and the subsidizing of journeys to foreign countries. Subsidized journeys of machine builders or students of the Gewerbe-Institut – not for pleasure but for working, as Beuth put it – to Britain, and from the 1840s to the main continental centres of machine building, to Belgium and to the Alsace were substantial not only for decisions in buying new implements. To work in or to visit a modern factory was the only way to learn the organisation of production in machine building, a subject not even mentioned in printed reports. Therefore it was almost compulsory for every machine maker before he started to produce locomotives to travel to Britain.[16]

To sum up: I tried to show a segment of the government transfer policy in Prussia in the 1830s and 1840s. The transfer of machine tools was and still is outside the focus of interest among historians of technology and even economic historians. The reason is that neither of them are asking how things are made; both are interested in the things, in the products. It seems to me, that politicians in charge of "Gewerbeförderung" like P. Ch. W. Beuth were not only very well informed about the products, not only deeply interested in buying machines and other implements but they also made efforts to learn how and by what kind of implements they were produced.

Therefore I would be very cautious to state, as many historians from Tipton to Pollard[17] have done, that state support in technology transfer failed, that government intervention did not support but hindered transfer. It was certainly a lump sum of the Prussian budget that Beuth spent on machine tools. But the support in overcoming the British barriers against the exportation of machinery, by supplying modern technology, by educating and training students of the mechanics' class on the best machine tools British factories produced, by enforcing the copying of modern technology and spreading the knowledge about it through subsidized publications as the "Verhandlungen", all this were small steps, not impeding but realizing the transfer and in the short run speeding up the industrialization of Germany. It seems to be very probable that the state support for acquiring machine tools from Britain also made its contribution to the improvements of the technical equipment in German machine-shops, taking place during the 1830s and 1840s. Whether technology transfer would have been more efficient without the intervention of the state seems to be a misplaced question in the real state of affairs in Prussia and could be answered by a séance of spiritualists rather than by historians.

Notes

1. This paper is based on my article "Der Technologietransfer für die Metallbearbeitung und die preußische Gewerbeförderung (1820-1850). In: F. Blaich (ed.), Die Rolle des Staates für die wirtschaftliche Entwicklung (Schriften des Vereins für Socialpolitik, N.F. Bd. 125), Berlin 1982, 99-142.
2. Myllyntaus, T., The Transfer of Electrical Technology to Finland, 1870-1930. Technology and Culture 32 (1991), 293.
3. Cameron, R.: The Diffusion of Technology as a Problem in Economic History. Economic Geography 51 (1975), 220.
4. Redlich F., Frühindustrielle Unternehmer und ihre Probleme im Lichte ihrer Selbstzeugnisse. In: Fischer W. (ed.), Wirtschafts- und Sozialgeschichtliche Probleme der frühen Industrialisierung, Berlin 1968, 343.
5. For the details see Paulinyi, A., Die Erfindung des Heißluftblasens in Schottland und seine Einführung in Mitteleuropa. Techikgeschichte 50 (1983), 13-20.
6. Jeremy, D. J., Damming the Flood: British Government Efforts to Check the Outflow of Technicians and Machinery 1780-1843. Business History Review 51 (1977), 28 ff.
7. Pollard, S., Peaceful Conquest. The Industrialization of Europe 1760-1870, Oxford 1981, 144.
8. Public Record Office, London; B.T. 1/310 No 19 (1835).
9. New Light on Richard Roberts, Textile Engineer 1789-1864. Transactions of the Newcomen Society 41 (1971), 27-44.
10. Pollard, S., ibid.
11. For the details see Mieck, I., Preußische Gewerbepolitik in Berlin 1806-1844, Berlin 1965.
12. Lundgreen, P., Bildung und Wirtschaftswachstum im Industrialisierungsprozeß des 19. Jahrhunderts, Berlin 1973, 142.
13. Gregory, O., A Treatise of Mechanics, Theoretical, Practical and Descriptive, London 1806, vol. 2, 471-75, plate 36; 'Pantologia'. A New Cyclopaedia etc., London 1813, vol. II, article: Turning lathe; Plate 172.
14. Beuth, P. Ch. W., Glasgow. Verhandlungen des Vereins zur Beförderung des Gewerbefleißes in Preußen 3 (1924), S. 124; Wolzogen, A. (ed.), Aus Schinkels Nachlaß. Reisetagebücher, Briefe und Aphorismen, Berlin 1863, 42 f., 80 f.
15. Wagenblass, H., Der Eisenbahnbau und das Wachstum der deutschen Eisen- und Maschinenbauindustrie, Stuttgart 1973, 88 ff. and 203 ff.
16. See Paulinyi, A. (1982), 126.
17. Tipton, F. B. Jr., Government Policy and Economic Development in Germany and Japan: A Sceptical Reevaluation. Journal of Economic History 41 (1981), 140 f.; Pollard, S., ibid., 163.

Sources for the Study of Industrial Espionage by Eighteenth century France

J. R. Harris

That one should be able to say anything at all about the sources for the practice of industrial espionage in the eighteenth century is sometimes a matter for surprise, not only to non-academics, but even to historians whose interests are outside the period or outside the history of technology. But even some closer to the theme, with an interest in eighteenth economic or even industrial history, are sometimes dubious that anything sufficiently connected or extensive is available to write a substantial paper, or even a book.

Naturally, industrial espionage was a secret activity at the time it was being carried on, and that might be thought to make it nearly invisible to the historian. There may be some truth in this as respects espionage within the industries of a single country, though there were of course very audible screams of rage from victim firms when they discovered the offence, as there are today. But when international industrial espionage was carried out the perpetrators generally wanted acclaim, financial reward, subsidies for manufactures which used the newly-gained secrets, offices under government in their own, the predator country, careful insertion of their achievement in the official files, or even honours, titles, and awards.

It took me some time to realise how rich a vein there was to mine. The discovery came about in this way. Over twenty years ago I thought of making a very limited comparison of French and English industry in the eighteenth century to see how far the technical lead of England over France derived from the earlier British reliance on mineral fuel and their conversion of nearly all their industry to coal-fuel technologies by the beginning of the eighteenth century. This I believed had very important consequences for both the long term direction, and the nature, of technological innovation in Britain, and was an important element in its comparative industrial lead. In turn, these leading-edge industries were something of a focus of French interest and imitation.

One outcome of this limited study¹ was my increasing interest in the type, and the transmission, of the craft skills involved in the new British technology, and in the ease of that transmission, both nationally and internationally.² Another outcome was a growing appreciation of French sources, particularly, but not exclusively, the National Archives, and the realization that they would allow set-piece comparative case-studies of some important industries in both countries. Successively I identified plate glass, steel and hardware. Once I got to work on these industries, the transfers of technology which were involved in each, and the place of coal-fuel technology in them, the problem proved to be not so much the exiguousness of the materials, but their bulk. It was the many pieces of the jig-saw which could be found, and the intricacies of fitting them together that was the difficulty, not the availability of so few pieces that they had little interrelation. The rather large articles which resulted could have easily become short books.³

The stimulus of an invitation to give a memorial lecture for a very celebrated historian of technology, L.T.C. Rolt, was very valuable. It made me try to look for a topic which might be at least a pale imitation of the lively writings with which he had been able to attract so many people to the study of the history of technology and of industrial archaeology, and his remarkable understanding of the place of individuals in it all. It occurred to me that much of the transfer of technology on which I had been working had been done by industrial espionage, and that this was an intensely human study full of remarkable individuals. As I rummaged in my notes it was obvious that there was even more reference to industrial espionage than I had remembered, and that once again the problem would be one of compression.⁴

This convinced me of other things. Given that I was studying the technological relationship between Europe's two leading industrial countries in the eighteenth century it became more and more evident that a focus on industrial espionage would be useful. Firstly it is the foreign technology which a nation feels is most important, the key technology, which it will make the most intense efforts to procure. For this it will encourage its nationals to run risks in the face of any protective legislation with which another nation employs to guard its technological skills and the people who possess them. Industrial espionage therefore highlights those technical areas in which the nation which practices it feels itself to be most deficient. Secondly, insofar as there is documentation about industrial espionage, it may tell us about the attitudes of businessmen, statesmen, scientists, and technologists in the country arranging the espionage towards new technical ideas and the plant in which they are embodied. It may show how these different groups evaluated the new foreign technology as compared to their own national practice, and how realistic their assessment of the possibilities of technical transfer were. Looking at the evidence of French archives, it was evident that the element of industrial espionage in technology transfer from Britain was high, and it became clear that a book about industrial espionage between the two countries would be a means of telling the story of technology transfer in an interesting way.

In this study it is French archives which are the major source, British archives a very minor one. Why? Because the French state did not abandon the belief that it had an important role to play in controlling, directing, advising, subsidizing and generally guiding industry before the Revolution, and it partly resumed that role later, though I have not gone on to tangle with the industrial policies of the Consulate, Empire and after. The system of regulation and encouragement, greatly extended under Colbert, was continued by later governments and their agencies, for instance by the Council and the Bureau of Commerce. The working of the latter has been well described by Professor H. T. Parker in his book which is focused on the Bureau in 1781,[5] but effectively outlines industrial administration from Colbert to that date.[6] From the 1740s to the late 1770s the Bureau was effectively headed by two Directors, the Trudaines, father and son. They were deeply devoted to the development of industry, but by no means convinced that regulation and subsidy were the panaceas; both were deeply interested in the English example and the attraction of English methods to France.[7] From 1779 until the early years of the Revolution their successor Tolozan, an able administrator if without their vision, continued to be interested in acquiring English men and methods, and supportive of industrial espionage. The Bureau had four Intendants of Commerce, each responsible for particular industries and a group of regions, and below them there was a numerous body of inspectors of manufactures, some times over 50 of them. Many had under-inspectors, while there were also some inspectors with specific limited industrial fields, travelling inspectors, and Inspectors-General with special powers. There were a number of these inspectors who had a particular interest in England, and some travelled there with official encouragement.[8]

Without going into further detail, it is clear that there must be, where accident has not intervened, a massive amount of paper work resulting from all these peoples' labours, often surviving in great detail and swollen by the processes of bureaucracy itself. Commonly at least part of the correspondence which arose on particular issues – for instance the local activities of English workmen in the French provinces, – will be recorded both in the National Archives and in those of modern Departments in which are preserved the records of the provincial intendancies of the Ancien Régime. The Intendants and their subordinates were supposed to back the policies and enforce the directives of the Bureau, and to provide information and opinion, so adding other layers to the file.

There are massive series of archives in the Archives Nationales which are devoted to these industrial materials, particularly in the categories F12 and F14, which cover the whole gamut of industry, including mining and metallurgy. But massive and intimidating as these are, they are by no means the only sources within the national archives. Another important group is that of the Marine. Here there are extensive papers on the acquisition of cannon-boring and coke-iron making techniques from Britain for naval artillery, a good deal on the copper sheathing of ships and associated cold-rolling techniques, and on the new British methods of approaching standardised production and standardised parts in the production of naval blocks and pulleys, and on British advances towards reliable marine chronometers. There is even some material on steam engines. This could also be followed up by looking into the archives of the naval ports themselves, but so far I have only looked at one port, L'Orient, which was of particular importance for the work of one major industrial spy, and for a temporary input from an early eighteenth century English industrial emigré. The army archives I have not looked at except vicariously, through the work of a research student.[9] She has worked through the exhaustive material, which happens to be in Army and not civil files, for the large number of English civilian "hostages" who were detained (not always unwillingly) in Napoleonic France. Here can be found over 75 commercial men and at least 316 manufacturers and artisans. Though the survey is not yet complete we can identify 177 people in the textile industry, including about 60 "mecaniciens", many of whom had been suborned to come to France either before 1793 or during the peace of 1802-3.

Naturally the national archives have their reflection in departmental archives, but, life being short, I have only looked at a few of these. Montpellier (Hérault) has among some other items a few papers submitted to the local Academy of Sciences on English technologies, Nevers an interesting collection on the introduction of English hardware methods at La Charité-sur-Loire, Rouen a good deal of background on discussions and comparisons about English and French technologies, though it is weak on the English workers in textiles and other industries. This is very disappointing, considering the known level of industrial activity in the region, and clearly this archive has been much disrupted at some time in the past.

There is a general opinion that French business archives are not as open to historians as British. This may be true, but I have been fortunate to see some useful ones, particularly those of Saint Gobain, where in the eighteenth century the plate glass company was raided for processes and personnel by the British, while itself making attempts to copy coal-using methods in glass furnaces and subsequently the English success in the steam-powered polishing and grinding of glass. The papers of the De Dietrich family, famous for centuries in metallurgical industry, can be consulted by arrangement in the national archives. Baron Frederic De Dietrich was a key technological adviser to government at the end of the Old Regime, and carried through an investigation into industrial fuel use just before the Revolution, with the idea of shifting industries to coal on the English pattern. Sadly it is mainly his political and social life which is reflected in the papers, but it is revealed that he was surveying the metal mines of Cornwall when the government recalled him for other duties.

Two other archives may be mentioned. Those of the French foreign ministry are useful. Industrial observers were sent over, sometimes with a clear espionage brief, and they often were offered the services of the French embassy in London. In some cases the Embassy was involved when there was a desire to recruit industrial specialists for French service, and there are some general observations on the improvements in English industry and the prospects of gaining them. Secondly, the Archives of the Conservatoire des Arts et Métiers are much neglected. Though the institution was founded in 1794 its archives carry a considerable number of files under textiles, metals, tools and other similar headings which often reflect back on the last years of the Old Regime and the first ones of the Revolution. Their fine collection of contemporary drawings of machines and even of factories is now being carefully catalogued and cared for.[10] Sadly for the industrial archaeologist, many of the English machines which were illicitly smuggled into France to be deposited in the Conservatoire's ancestor, the Hotel de Mortagne, and after 1794 in the Conservatoire itself (though carefully acquisitioned at the time), seem to have been sacrificed by space-hungry conservators in the last century. However there are many mentions of English artisans and technologists in the archives.

For a moment let us look over the Channel. How far do British archives give help to the historian looking for foreign industrial intelligence and espionage? From a national aspect very little. The British industrialist was left to find his way to wealth and respect or to a commission in bankruptcy with very little attention from the state, if we except taxes, customs and excise.[11] With the exception of a few instances in a couple of periods the situation is in great contrast with France. Neither the papers nor the Journal of the Commissioners for Trade and the Plantations (apart from the cases mentioned later) nor even those of the later Board of Trade, nor the State Papers Domestic, nor their successor the Home Office papers have much to offer. State Papers France have a few high points at the beginning of the century, but die away thereafter. Of course much depends on the quality of the indexing of some of the national series, and this varies. I have been able by target spotting to find some nice items in Admiralty records at the National Maritime Museum, but there are hundreds of volumes of correspondence indexed only alphabetically by correspondent, while the correspondence with the officials of the yards at the different ports awaits calendering. Though, as we will see, legislation was in force against the unwanted transfer of technology, nothing has been done as yet to check litigation in the major courts for evidence of it, but this neglect is true for almost all of British industrial history.[12] The existence of provincial newspapers especially after the middle of the century throws up a few instances of espionage or suborning.

Business archives are a help in some cases. It has long been known that industrial espionage and the theft or loss of workers features in the Boulton and Watt papers, and this was long ago discussed by Eric Robinson.[13] There is some more ore in that mine. But Boulton was a perfect weathercock on this issue, turning wherever his momentary personal interest dictated. Far more consistent was Josiah Wedgwood, who was steadily and sternly against the theft of methods and men, and produced a powerful pamphlet to dissuade his workers from desertion abroad.[14] There is some good evidence here.

At this point let us mention the great exception to the dearth of material in British archives. This brief period of plenty begins in 1718 and runs over into the early twenties. The impetus was a great scheme organized by John Law, the Scottish financier, who came to dominate the monetary scene in France, was associated with the famous bubbles of the Mississipi Company and the Royal Bank, and was briefly first minister of France. His plan, which I have recently described, was to bring a large number of British workers to France in those trades where Britain had the leading edge. There were probably over 200 men involved in horology, iron founding, textiles, steel, glass, shipbuilding and other skills, with concentration on Normandy, Paris, Versailles and Lorient. This vast operation led to a panic in Britain, and the passing of rapid legislation which was directed at preventing the exit of skilled artisans, securing the return of the men who were already abroad, and placing very heavy penalties on suborners. There is no question that the essential element in the transfer of technology at this time was the movement of the skilled worker, often with few or no tools. Thus the legislation concentrated on the central means of industrial espionage.[15] It was only from 1750 that the law, in a patchy way, began to include the export of machines in certain industries in its ban, with Acts of 1750, 1774, and 1781 against textile machinery exports, and those of 1785 and subsequent years against the export of the equipment of the Birmingham trades, of the metal industries and some others.

Much of the transfer of technology thus became illegal. In archive terms the 1718-20 Law scheme causes an unusual source situation. Though not written up until very recently, there is a mass of British documentation for these years; the normal sources of Parliamentary petitions and the steps towards legislation; many representations to the Commissioners of Trade; correspondence from the Embassy in France; approaches to the relevant Secretary of State; many items in State Papers Domestic. By contrast the French national and departmental archives know practically nothing of this extremely important operation, and it has remained unknown to past and present historians in that country. The dislocation of archives after the disgrace and flight of Law, and the extent to which the scheme was very much his own, may explain this. Later in the century, especially from the late 1740s, it is the plentiful archives of the dirigiste France which provide most of the evidence, and it is from the French State, or via the State, that the efforts to transfer technology, in large part by industrial espionage, derive.

Many of these are fascinating, but we can only mention a few in passing. John Kay, of fly-shuttle fame, passed much of his life in France.[16] John Holker, a Jacobite officer, was brought out of the French army, and from the early 1750s set up very successful textile concerns in Normandy. He became a close adviser to Government on industry, and was appointed Inspector General of Foreign Manufactures between the mid-1750s and the early 1780s, with the job of bringing British methods to France by industrial espionage.[17] As well as textile methods he brought sulphuric acid production. Michael Alcock brought Birmingham hardware methods in the 1750s, and endeavoured to seduce metal workers from his own locality,[18] as did others. Gabriel Jars made the famous tour of Britain which was a main basis of his celebrated work, the Metallurgical Travels; the records make it clear that he was supposed to spy.[19] With Government support De Saudray made a special effort to bring more Birmingham methods in the 1770s. There was a long continuous period of espionage in the mid-80s by Le Turc, in which the biggest activities were the very successful transfer of naval block and pulley making and the shipment of stocking frames – even a whole workshop at once – to France.[20] Then from the late 1770s and the 1780s, originally partly connected with Holker, came the flood of men introducing the new textile machine technologies.

On most of this British archives are silent. We know that spies were pursued, suborners sometimes arrested and jailed, and there were newspaper references to acts of espionage. In the mid-1780s, partly related to the Anglo-French commercial treaty of 1786, and the temporary importance of local commercial committees in the great industrial towns, there was a noticeable recrudescence of reported cases, and men like Josiah Wedgwood and Samuel Garbett of Birmingham were prominent as supporters of the laws and as suppliers of information to Government. We have to remember that contemporary Britain was not a police state, and was virtually without a police force.

Again the law was not necessarily backed by industrialists. The works of some of them, like those of the Wilkinson ironmasters and of Thomas Williams the copper tycoon, were virtually open to all comers. Boulton and Watt vacillated hugely, but they tried to export their steam engines, even in wartime, and got a monopoly privilege for France, though few engines actually went abroad. There was never any opposition by Government, or public, to the export of any steam engine parts or to the travels of erectors. There was a clash between the idea of exporting those machines which could be regarded as having some degree of scientific content, and were part of the Enlightenment, and the need to protect home industries.[21]

There was also a considerable difference between the attitudes taken to persons of different social and economic classes who wished to travel abroad to practice their industrial skills. There was virtually no opposition to men of the employing class (even if they were on the lowest rungs of the capitalist ladder) who travelled abroad to work on their own behalf or with foreign partners and associates. On the other hand the mere skilled workman who had been suborned would be prevented without compunction from emigrating. Marchant de la Houlière quite accurately presaged the ease with which one of the Wilkinson brothers might emigrate to France and serve the French government provided he went in peacetime: "he could be allowed, by virtue of his rights as an Englishman, to follow up the business upon which his fortune depended." [22] This paragraph, and the preceding ones, provide some of the reasons why British archives on industrial espionage are much inferior to French.

While no single monograph has yet been published on the subject of eighteenth century industrial espionage, there is of course a substantial reference to it in some important books old and new, and some very interesting papers have appeared. Some of both are cited here without any attempt at bibliographic completeness.[23] Finally, perhaps, it may be worth giving some examples of the remarkable contemporary comments and descriptions which can be found in the sources; fittingly they are all from French ones.

In advancing the services which he could provide to France as its industrial spymaster Holker wrote "If one proposes to bring to France foreign skills and principally those of England, where industry has made more progress than anywhere else, one can first use [me] to establish and keep up a secret correspondence with England to get thence certainly and quickly all the models of machines, and the samples and tools one needs." The best way to evade the law was to import these things via Holland, as Dutch trade was less closely supervised by the Customs service. English workers should be recruited by sending over an English workman already in France who could accompany them to their French destination, thus reducing the risk of their running away once suborned and avoiding language problems. It was best to avoid culture shock by recruiting Catholics, or by obtaining young men who could be given dowries to marry French girls and pensions for conversion to Catholicism.[24]

While their ability to penetrate English industry was exceptional, in that they had Perier with them who was dealing with Boulton and Watt on engine matters, and their visits were particularly directed to the iron and copper works of the Wilkinsons and Thomas Williams, who were unusually open with visitors and often unconcerned about secrecy, the view of the artillery officers De Givry and De Wendel (the latter also an ironmaster) [25] is a remarkable one. "We found that there was nothing difficult in getting a good view of English manufactures, one needs to know the language with facility, not show any curiosity, and wait till the hour when punch is served to instruct oneself and acquire the confidence of the manufacturers and their foremen, one must avoid recommendations from Ministers and Lords which will do little good and make contact with some of the principal industrialists who can open every door ... young men are little suited to such a mission ... to look at things usefully it is important to have at least some idea about machines, because one does not take a step without seeing them, which all tends to abridge the process of manufacture."

The essential place of the skilled worker in transferring technology is frequently alluded to. "The arts never pass by writing from country to another, eye and practice can alone train men in these activities",[26] or again "Even written materials are a very feeble aid to the mechanical arts, they cannot impart the skill, one can have it by work alone".[27] Finally the recognition by an industrial spy that industrial revolution (at least as historians of technology can recognize it) had arrived in Britain. He "saw with dismay that a revolution in the mechanical arts, the real precursor, the true and principal cause of political revolutions [he meant in the international balance of power] was developing in a manner frightening to the whole of Europe, and particularly to France, which would receive the severest blow from it".[28]

Notes

1. J.R. Harris, *Industry and Technology in the Eighteenth Century, Britain and France*, (inaugural lecture, University of Birmingham, 1972). See also "The Rise of Coal Technology", *Scientific American* 231 No. 2. (1974).
2. "Skills, Coal and British Industry in the Eighteenth Century", *History* 61, (1976).
3. "Saint-Gobain and Ravenhead" in B.M. Ratcliffe (ed.) *Great Britain and her World 1750-1914. Essays in honour of W.O. Henderson* (Manchester 1975); "Attempts to Transfer English Steel Techniques to France in the Eighteenth Century" in S. Marriner (ed.) *Business and Businessmen* (Liverpool 1978); "Michael Alcock and the Transfer of Birmingham Technology to France in the Eighteenth Century". *Journal of European Economic History* Vol. 15, No. 1, (1986).
4. "Industrial Espionage in the Eighteenth Century", The Rolt Memorial Lecture, 1984, *Industrial Archaeology Review*, Vol. VII, No. 2. (Spring 1985). The papers listed in this and the previous note have just been collected in *Essays in Industry and Technology in the Eighteenth Century* (Variorum 1992).
5. Harold T. Parker, *The Bureau of Commerce in 1781 and its Policies With Respect to French Industry.* Carolina Academic Press (Durham N.C. 1979). Parker not only gives the administrative practice of the Bureau but sensitively and sympathetically recreates the whole atmosphere within which industry was supervised and stimulated.

6. There was a good deal of revision of, and tinkering with, the system of industrial administration and the duties and remit of the main functionaries. See, in addition to Parker, P. Bonnassieux and E. Lelong, *Conseil de Commerce et Bureau de Commerce 1700-1791* (Paris, Imprimerie Nationale, 1900), Introduction, p.x seq. and T. J. Schaefer, *The French Council of Commerce 1700-1715* (Columbus Ohio, 1983).
7. There seems to be no good recent account of the industrial policy of the Trudaines, though Parker has some very useful passages on them, e.g. 101-3, 134-5, 180-2. The remarkable co-operation of the Trudaines with John Holker in obtaining and using English technology is covered in André Rémond, *John Holker, Manufacturier et Grand Fonctionnaire en France au XVIIIe Siècle 1719-1786* (Paris 1946); there are older writings by E. Chouiller, *Les Trudaines* (Arcis sur Aube 1884) and a paper of the same title of 1883 which derives (like the former) from the *Révue de Champagne et de Brie,* T.5. There is an "éloge" of the elder Trudaine by his son in the *Histoire de l'Académie Royale des Sciences,* premiere partie, MDCCLXXVII. (B.N. R14902) Tolozan's work will receive attention in Professor Parker's forthcoming continuation volume.
8. F. Bacquié, *Les Inspecteurs des Manufactures sous L'Ancien Régime 1669-1792* (Paris 1927), though now somewhat outdated in approach, remains very valuable. Inspectors and industrial commissioners who visited England include Morel, Ticquet, Jubié, Gournay, Hellot, Jars, De Dietrich and Faujas de Saint-Fond.
9. Mme. Margaret Audin, "British Hostages in Napoleonic France. The Evidence with particular reference to Manufacturers and Artisans" (M.Soc.Sci. thesis Birmingham 1987).
10. I have to thank Dominique Ferriot, Director of the Musée National des Techniques, for excellent access to these archives at a time when the museum was engaged on planning its complete reorganization and re-development.
11. "But the greatest benefit which the government confers, the greatest which it can confer in any state, is that of doing nothing at all: in the whole of this country there are no regulations, nothing is forbidden, and there is no need of particular encouragement in any one department of the state. The greatest source of encouragement in any class of men whatever is their own personal interest... Everyone can sell his goods to whom he likes and at the highest price he can get; traders go anywhere where they think they can get a profit; there is no interference whatever with their business, and in the eyes of an impartial traveller, England has the appearance of being a hundred times richer than France". This contrast was made by Francois de la Rochefoucauld. *A Frenchman in England, 1784* (ed. J. Marchand, trans. S. C. Roberts, Cambridge 1933).
12. Janet Butler, in a current study of John Wilkinson the ironmaster, is a remarkable exception to the general failure to use this source.
13. Eric Robinson, "The International Exchange of Men and Machines", *Business History* 1 (1958) 3-15. There is a later version in E. Robinson and A. E. Musson, *Science and Technology in the Industrial Revolution* (Manchester 1969) ch. VI.
14. See his *An Address to the Workmen in the Pottery on the Subject of Entering into the Service of Foreign Manufacturers.* (Newcastle, Staffs 1783).
15. An account of this important episode is now available as "The First British Measures against Industrial Espionage", in *Industry and Finance in Early Modern History, Essays presented to George Hammersley,* ed. I. Blanchard, A. Goodman and J. Newman (Stuttgart 1992).
16. A. P. Wadsworth and J. De L. Mann, *The Cotton Trade and Industrial Lancashire,* Ch. XII, 449-465.
17. For his biography, A. Rémond, op.cit.
18. See above, note 3.

19. Archives Nationales, F12, 1311.
20. A paper on "A French Industrial Spy: Le Turc in England in the 1780s" is to appear in a festschrift for François Crouzet.
21. See n.4 above.
22. W. H. Chaloner (ed.) *Marchant de la Houliere's Report*, reprinted from *Edgar Allen News* (1948-9).
23. Of particular importance are W. O. Henderson, *Britain and Industrial Europe 1750-1870* (Liverpool 1954), and, especially for France, C. Ballot, *L'Introduction du machinisme dans l'industrie française* (Lille, Paris 1923) and the thesis by L. de Bellefonds, "Les Techniciens Anglais dans L'Industrie Française au XVIIIe Siècle" (Univ. of Paris 1971). Important papers are Eric Robinson's (above note 13), David Jeremy's "Damming the Flood; British Governmental Efforts to check the Outflow of Technicians and Machinery", *Business History Review* 51 (1977), and Peter Mathias's "Skill and the Diffusion of Innovations from Britain in the Eighteenth Century" *Trans Royal Historical Society*, 25 (1975). Harris (art.cit note 4) mentions the Swedish official observers in eighteenth century England and some of the sources for them, which provide an interesting comparison with the French. Kristine Bruland's edited volume carries some papers of great relevance for Scandinavia (*Technology Transfer and Scandinavian Industrialisation*, Oxford/New York, 1992) especially R. P. Amdam's "Industrial Espionage and the Transfer of Technology to the Early Norwegian Glass Industry" 73-93. Some comparison is possible with Austria due to H. Frendenberger, "Technologischer Wandel im 18. Jahrhundert", *Wolfenbuttel Forschungen, Band 14* (Wolfenbuttel 1981). The present writer hopes to complete in 1993 a book on eighteenth century French industrial espionage in England.
24. P. Boissonade, "Trois Mémoires relatifs à l'amelioration des manufactures de France sous l'administration des Trudaines (1754)" in *Revue d'histoire économique et sociale* (1914-18), II, Mémoire tendant a multiplier et perfectionner les fabriques de France", 68 seq.
25. Archives Nationales, T 591 4 and 5.
26. Boissonade, *op.cit.*
27. Archives Nationales, F12 677c.
28. Ibid. (Letter of Le Turc to Citoyen Ministre, 5 Fructidor Year 7).

The Royal Danish-Norwegian Dockyard. Innovation, Espionage and Centre of Technology

Frank Allan Rasmussen

Introduction

This paper will sketch the technological development of a state owned military establishment during the 18th century. It is my aim to try to identify some of the causes for its outstanding position as the leading technological force of the dual Monarchy of Denmark-Norway.

The purpose of this paper is to examine some of the changes that took place in the Royal Naval Dockyard in Copenhagen. It will be argued that this institution succeeded in keeping abreast of the practical and theoretical development in Europe, not only in the field of naval architecture and shipbuilding, but also within the wider field of technology as such.

I shall present evidence showing how successive administrations of the naval dockyard were capable of maintaining a front position in a continued process of modernization for more than 150 years, not only in terms of innovation but also in terms of implementation of new technology.

The professionalization of the naval officers was the distinguishing trait of all, even if compared with other European countries. The constituent elements in this process were the founding of institutions of practical and theoretical education, intensive espionage and technical transfer from foreign industrial centres.[1]

Tasks and doctrines

Throughout the 18th century, the navy of the twin-monarchy had two main objectives. The first task was to secure the internal lines of communication and to protect the countries from a seaborne invasion. This together with the fact that

Denmark and Norway were separated by the sea, made the protection of the trade and other transports the primary assignment of the navy.

You must bear in mind that Denmark was a colonial power with trading posts scattered over enormous distances. Towards the south, in the Westindies, Westafrica and even in the Indian Ocean. Towards the north, the immense empire of Greenland and Island. The control over the trade in these remote parts of the monarchy was based solely on the navy.

Denmark had no territorial aspirations, but throughout the period the second task was to preserve the independence and integrity of the twin-monarchy. This made Sweden's policy of aggrandizement, and the rising naval power of the Russian Empire the only serious threats. Therefore the main doctrine of the Danish Admiralty was to have a fleet that could fight the Swedish and Russian navies at least on equal terms.

To this must be added the need of a navy sufficiently large and well equipped to constitute a potential alliance partner in the European Theatre of War.

The strategic position of Denmark played a major role. For centuries the monarchy had been in control of the entrance to the Baltic and the access to its enormous resources. In order to maintain this controlling position, it was necessary to keep pace with technological development.[2]

How was this done? – Which were the means brought into action? – And how did the Admiralty acquire the intelligence so badly needed?

Shipbuilding and shipyards

The Danish man of war of the 17th century was built by constructors from England, Scotland and the Netherlands.

Until the beginning af the 18th century, shipbuilding was carried out by way of tradition. It was a handicraft, and its technique was characterized by conservatism and secrecy. The knowledge was based on experience and a set of rules passed on from father to son, or from master shipwright to journeymen.

These foreign constructors were international experts who offered the secrets of their trade to the governments who were willing to pay their price. But at the beginning of the last decade of the 17th century this practice was changed by the Danish Admiralty, partly because these foreign shipwrights did not want to share their knowledge with Danish shipbuilders, partly because these masters often broke the seal of secrecy by leaking information about the Danish-Norwegian Navy to their native countries.

One of the last foreign master shipwrights, Francis Sheldon (1612-1692) was dismissed in 1690. He returned to England, but his descendants served in Sweden for more than six generations.

In Denmark the family tradition was abolished by the turn of the century, and the navy was almost unique in recruiting its naval architects from the officer corps.

From the late 1690s, young naval officers proficient in mathematics and naval architecture were sent abroad to study for several years. These studies frequently crossed the limit of industrial espionage.

I shall return to this point later in this paper.

It is remarkable how, at the turn of the century, the Admiralty managed to create a national construction corps in accordance with the predominant doctrines of naval warfare and the technological standards of their time. This does not imply, however, that there was complete agreement as to the question of how to build a man of war. Throughout the period there were continuous disputes between theoretically orientated constructors and practical shipbuilders: Should the man of war be a good sailing ship or a good warship? Two qualities which were difficult to combine in the relatively small ships of the Danish-Norwegian navy.

In spite of theoretical debates among leading naval officers, no final answer was found to these technical problems. The issue can best be understood by reducing it to a general disagreement amongst senior naval officers, as to whether French or English methods of shipbuilding should be practiced.

The disagreement did not originate from intensive studies in the literature on naval architecture; it was simply a question of different personal experiences, having been gained by officers in studying other navies in Europe.

From the turn of the century constructional drawings testify the prevailing French influence on Danish shipbuilding. The admired qualities of the French man of war, speed and firepower, were generally acknowledged.[3]

At the same time the Admiralty made great efforts to change the organization of shipbuilding. In the 15th and 16th centuries naval vessels were built on locations where resources and man power were available, i.e. along the Baltic coast and in Norway so affluent in timber.

But in 1690 existing facilities, in the heart of Copenhagen, were greatly improved and enlarged. From now on, all ships were built in Copenhagen. In other words a centralization took place in the capital, near King, government and Admiralty, thus providing a better opportunity to control resources and labour.

The naval dockyard became the largest place of employment in the twin-monarchy, with an average number of workers, officers and administrators of 5.500, approximately 15 per cent of the male population of the capital.

An average of two ships of the line and one frigate, together with some smaller vessels were built each year.

But it was not only ships that were produced in the dockyards. In accordance with prevailing mercantilistic ideas the supreme command of the navy made an attempt towards self-sufficiency with regard to all equipment necessary for a modern navy, viz. forging of anchors, weaving of cloth and sails, casting of canons, manufacture of gunpowder, ropemaking, etc.[4]

In a sense, the dockyard can be used as an indicator of technological competence in the twin-monarchy.

Institutions of education

I have already mentioned two of the major changes that took place at the turn of the century: nationalization and centralization of shipbuilding. To this must be added the founding of institutions of practical and theoretical education.

In 1701 the Admiralty founded a Naval Academy. In this institution young midshipmen were given instruction in mathematics, drawing, construction, navigation, fortification, fencing, dancing and religion, together with lessons in English and French.

As a supplement to theoretical education, the young pupils had to take part in training cruises during summer time. The teaching staff was recruited among naval officers and professors of the University of Copenhagen.

The objective of this institution was to provide not only naval architects but scientifically trained naval officers for the fleet.[5]

The pattern was taken over from French institutions as laid down by the reformer J.-B. Colbert (1619-1683). During the 1680s he had established the "Ecole Flottants", and if we compare the syllabus of the two institutions, the only difference is the language teaching by the Danish Naval Academy.[6]

In 1739 another new institution, the Construction Committee, was established. The basic aim of the committee was to assess and approve the constructional drawings provided by the chief constructor to the navy. Upon completion of the man of war the committee had to control that the ship had been built according to the drawings, approved by the king.

Eventually the Construction Committee became the centre of technological knowledge of the naval dockyard and it became the leading expert organ on all technical questions within the twin-monarchy.

As part of its effort to encourage young naval officers to improve their proficiency in mathematics and shipbuilding the Construction Committee also introduced a system of auscultation, that is, young naval officers were admitted to observe the meetings of the committee.

To support the construction of ships the Admiralty had established a collection of ship-models dating back to the end of the 17th century. This collection comprised scale models of nearly all ships built at the naval dockyard together with the original constructional drawings. This unique material enabled naval constructors as well as the committee to make very useful comparisons between new and old constructions, and to compare their individual qualities.

Another technique applied by the Admiralty was a test system whereby old well known ships were tested against new ones. These test results were recorded and filed for later use.[7]

Once again this pattern was borrowed from France. In the 1670s Colbert had established the "Conseil du Construction" which had a similar competence.[8]

In 1757 a school of naval construction was established. It was intended as a practical workshop. However, the students would have both theoretical and practical experience in making sheer draughts, models and laying off lines. At this institution the most up to date theoretical knowledge of science and shipbuilding was taught by the chief constructor of the navy.

Gradually apprentices and younger craftsmen became students of naval architecture, yet another indication of the ongoing process of professionalization.[9]

During this period Europe witnessed a boom in the publication of works of reference on naval architecture. In the Netherlands the great epoch of trade and shipbuilding was brought to an end by the publication of the works of Nicolaes Witsen, in 1671, and his contemporary colleague Cornelis van Yk, in 1697.

In Great Britain, normally not known for its publications on shipbuilding, two prints were made shortly after the turn of the century. The first one was published by John Hardingham in 1707, and the second one by William Sutherland three years later.

In France the jesuit Paul Hoste published his work in 1697, Jean Bernouilli in 1717 and later the two famous naval architects Pierre Bouguer and Duhamel du Monceau in 1746 and 1757 respectively. The list could be continued.[10]

But why this sudden and overwhelming interest in the art of shipbuilding?

One explanation might be, that scientists were beginning to look for new fields for the testing of their theories, and in the field of shipbuilding they found a challenging object. Secondly the naval-powers of Europe were beginning to understand the necessity of setting shipbuilding into a more rational framework.

The expansion of colonial trade and the incipient shortage of wood were other important factors. Shortage of wood and improvement of naval artillery forced new challenges upon the field of naval architecture.

But to the scientist the challenge seemed to be too great. A naval vessel is a large, dynamically loaded body moving on an uneasy element under the influence of buoyancy, wind and waves.

The few and scattered scientific experiments did not bring about any solutions to problems such as stability, speed, lightness, seaworthiness, durability and firepower. These qualities could only be achieved at the expense of one another. Shipbuilders had to make priorities in accordance with specifications from their superior officers. It is, perhaps, a typical feature of the Enlightenment that the ongoing discussion about the proper principles of ship construction made the leading seapowers turn their attention towards scientific services. Government officials enjoyed an almost blind faith in science. They believed that scholars were in possession of the key to improving shipbuilding along theoretical lines. They proved to be wrong. Scientific shipbuilding never became a success.

Let me sum up: the military industrial complex in Denmark-Norway had a sort of monopoly on technical and theoretical know-how. The institutions created during this period, the Naval Academy, the Construction Committee and the School of Naval Architecture played an important part in the ongoing professionalization of naval officers and constructors. The education system was adjusted to the general technological development and corresponding adjustments were made in regard to the organization inside the dockyard.

Over a period of about 30 years the Construction Committee became the institution where expertise on naval architecture and shipbuilding was accumulated, both from at home and abroad.

Almost certainly these initiatives made a greater contribution to improving shipbuilding than scientific works of reference on naval architecture.

Industrial espionage

To improve the education at the Naval Academy promising young naval officers were offered a supplementary dimension abroad. Often these activities would assume the character of industrial espionage. At the beginning of the 18th century these activities were a matter of routine. The Admiralty took the initiative, made up the plans for the journeys, and paid the costs.

In France we find a similar pattern. Colbert also urged his naval officers and agents to travel abroad especially to England and the Netherlands, and when these activities were made illegal in Britain during the second decade of 18th century, they changed to become an art of industrial espionage, as we have already learned from professor John R. Harris' most interesting paper.

These journeys were dangerous and punishment in case of unmasking was severe.

Let me illustrate these activities by a few examples based on source material in Danish and foreign archives.[11]

In 1757 Frederik Michael Krabbe (1725-1796) was appointed chief constructor to the navy. Krabbe had attended the Academy for several years and for a period he was seated as an auscultant in the Construction Committee.

In 1752 he received his instructions from the king to go to England, France, Italy and the Netherlands. Krabbe went to England the same year and shortly after his arrival the Admiralty received the first reports from his visits to the naval dockyards in Deptford and Woolwich.

Krabbe concealed his status as a naval officer by pretending to be an ordinary Danish traveller. But soon Krabbe was facing his first difficulties. He learnt the difference between the Danish fashion of dress and the English one. He therefore had to see a tailor who tried to make him look like an English gentlemen. In spite of this investment Krabbe's true identity was revealed at the naval dockyards of Portsmouth and he was expelled.

One of his many reports to the Admiralty describes his troubles and his fears to be identified and arrested. Krabbe spent more than a year in Britain and during that period he visited the naval dockyards of Deptford, Woolwich, Portsmouth, Chatham, Plymouth and Sheerness. Obviously he was very successful. His letters prove that he collected a lot of intelligence on shipbuilding.

In the spring of 1754 he went to France carrying a sea-chest full of notes and drawings stolen from dockyards in Britain. In France surveillance of foreign travellers was more rigorous.

His reports from the dockyard at Brest show that he had to be extremely cautious. Nobody noticed his presence at the naval dockyards, but he did not even dare to speak to shipbuilders or to use his notebook and pencil.

At the dockyard of Rochefort he had to employ new methods. He bribed the master shipwrights to bring him drawings which he copied in nocturnal solitude in order that the material could be returned the following morning.

Krabbe also went to Paris to see Pierre Bouguer (1698-1758) and Duhamel du Monceau (1700-1782), two of the most outstanding naval architects of the period, and he had the opportunity of getting to know their pioneer works.[12]

After his visit to France, Krabbe went to Italy visiting the naval yards of Pisa, Venice, Neaples, Arcone and Civitavecchia, and making a private excursion to Rome.

He had no high opinion on the state of shipbuilding in Italy and he described the natives as corrupt, ineffective, and lazy.

His stay in Italy made no contribution worth mentioning so he left for the last sojourn of his itinerary, the Netherlands.

Krabbe arrived in Amsterdam in August 1756, but finding no relevant information in the Netherlands, he decided to return to Denmark.

He had now spent more than four years abroad. Back home he was ordered to report to the board of the Admiralty. Soon after he was appointed chief constructor to the navy. The intelligence he had collected by travelling and spying proved most useful for designing and building a large number of fine ships.

His successor from 1772 was Henrik Gerner (1742-1787), who had also passed through the examination system and been spying abroad.

His work at the naval dockyard and for private yards gave him a reputation as one of the most talented naval architects ever in the twin-monarchy. Gerner was a contemporary of the Swedish naval architect F. H. Chapman (1721-1808).[13]

Gerner was familiar with the theoretical works of Newton, Leibnitz and Euler and he moved among the professors of the university. He was a member of the Royal Danish Academy of Sciences and Letters and a chairman of the Board of Agriculture.

In 1764 Gerner made a secret expedition to Sweden and Russia. He carried out espionage in the naval dockyards of Carlscrona, Stockholm, Reval and Cronstad, bringing back to the Admiralty reports on shipbuilding, timber, forging of anchors and casting of canons.

In 1768 he followed the footsteps of Krabbe. In England he played a double role. One as a secret agent, the other as a Danish scientist. He was appointed a member of the Society for the Encouragement of Arts, Sciences, Manufacture and Commerce, and among leading English scientists he gained a reputation of being a most promising young mathematician. He spent more than four years abroad, and at the age of 30 he was appointed chief constructor.

One of Gerner's successors was F.C.H. Hohlenberg (1765-1804). He had great international experience and his innovative designs had a major influence on shipbuilding, not only in Denmark, but in Europe. Basically his design reduced the size of the stern. The objective was to provide the greatest possible means of defence within the smallest possible target presented to the enemy. This together with his rationalization and standardization gave him an international reputation.

At the turn of the century the Danish Admiralty had one of the most modern shipyards in Europe and indeed a very professional corps of naval architects and shipwrights. The Danish man of war was recognized all over Europe. This position was achieved because successive administrations of the naval dockyard recognized the importance of investment in education and did not refrain from the suspicious trade of industrial espionage.

Following the terror bombardment of Copenhagen in 1807 and the Danish capitulation, the golden era of sails was brought to an end.

On a cold October morning of 1807 the British fleet left Copenhagen capturing the Danish fleet, together with more than 250 transports of material and equipment stolen from workshops and naval stores.

In this way the English killed two birds with one stone. Not only had they brought an end to Danish supremacy in the Baltic, they had also acquired unrestricted access to the know-how gained by Danish naval architects throughout generations.

Back in England, the Danish naval vessels were disassembled, measured and complete sets of constructional drawings were made. On this basis a series of eminent warships was built, serving the British fleet for years. It was no more than a poor consolation that this enterprise was an indirect proof of the high technological standard of the Royal Danish-Norwegian Dockyard.[14]

The case of the steam engine

The second case is an example of industrial espionage, too, but outside the realm of shipbuilding proper. It offers an illustration of the problems connected with the transfer and implementation of advanced technology.[15]

In 1790 the first steam engine was erected in the naval dockyard.

Three years earlier a young Danish naval officer had been sent on a secret mission to England. The aim was to procure information on steam engines and mechanical forgehammers.

The British "Tools Act" forbidding the exportation of machinery and know-how, made this a very difficult and dangerous task. Shortly after his arrival he got in contact with Andrew Mitchell, a Scottish engineer who promised him to make a model and drawing of the complete machinery.

The plans were accepted by the Admiralty in Copenhagen and soon after parts of a steam engine and a forgehammer were produced separately in different English workshops and smuggled out on board various ships, together with Mitchell and Alexander Young, a workman.

Following a lot of technical problems related to the installation of the machinery, the steam engine was ready to be tested in the spring of 1790. The original budget had then been exceeded by more than 600 per cent, and still it appeared that the problems of operating the steam engine had not been solved. In 1801 the Admiralty lost patience; Mitchell and Young were discharged and the steam engine demolished.[16]

As a consequence of the growing number of ships in the Danish navy an efficient production of anchors was badly needed.

This created a bottleneck situation which the Admiralty tried to solve through illegal transfer of new technology.

This is a classical example of technology transfer, and only a few years later the story repeated itself.

In 1803 the Danish government ordered an agent to visit Boulton & Watt's advanced industrial centre in Birmingham. He carried instructions to try to persuade Boulton to deliver a steam driven coining press and other machinery to the Royal Danish Mint and two steam engines to the naval dockyard.

In may 1804 a contract with Boulton & Watt was signed and shortly afterwards the House of Commons passed a bill approving the exportation of a complete mint to the Kingdom of Denmark. However the Act of Parliament does not mention the extra steam engines and machine tools for the naval dockyard.

In spite of the fact that Denmark and England were in a state of war, Boulton succeeded in smuggling out the extra machinery under cover of the otherwise legalized exportation. Being contraband of war, the steam engines and the machine tools obviously violated the Tools Act.

To make a long story short: the machinery did not solve the problems at the Mint nor those of the naval dockyard. Meanwhile the Admiralty was struck by unpredictable events. As I have already mentioned, the navy lost all its ships in 1807, and consequently the demand for the forging of anchors had disappeared. The Admiralty was in possession of a complete steamdriven machinery, advanced technology that was no longer needed.

The Danish government had made a strong effort to accomplish this technology transfer. The over all reason had been to become independent of supplies from abroad, according to prevailing mercantilist ideology.

In the governmental offices civil servants had put too much confidence in the potential of new technology, but had severely underestimated the difficulties related to technology transfer. To a certain extent they had been lead by ideology, not by existing realities.

But even if the technology transferred did not match the technological level of the recipient society, it had some side effects. The attention was drawn to the fact that Danish craftsmen and "engineers" lacked technical skill, and that the only solution to this problem was training and investment in modern machine tools.[17]

A new epoch

After the catastrophy of 1807, and following the war against England (1807-14) the navy had to rebuild the fleet and at the same time to procure technical expertise in the new fields of maritime technology.

The well known method of sending naval officers abroad was still useful, but now together with semi-governmental agents.

One must be aware that there were only a few firmly established mechanical industries outside the naval dockyard and no institutions had yet been established to take care of the need for civil technical education.

Only a handful of Danes met the requirements necessary to operate steam engines and machine tools.[18]

In 1842 a young Danish naval officer was sent to England to study steam engines and machine tools. He was ordered to suborn an engineer to come to Denmark. The naval officer was highly successful. He met the technically gifted William Wain (1819-1882). In 1851 he was appointed leader of the new mechanical workshop of the naval dockyard.

In 1852 he obtained citizenship and 10 years later he was appointed assistant director of the naval dockyard. Under the leadership of William Wain, and employing his personal contacts in England, he acquired a series of new and advanced machine tools.

In 1865 Wain became a partner of the private dockyard and mechanical firm "Burmeister & Baumgarten", from now on called "Burmeister & Wain".

The second half of the 19th century was characterized by the fact that both knowledge and technology had to be purchased and transferred from abroad.

An intense and fruitful cooperation with leading industrial establishments was initiated during this period, such as Krupp, Siemens, Telefunken and Zeiss in Germany, Thornycroft, Maudslay and Whitehead in the U.K., Creusot Forges et Chantiers and Donnet Leveque in France, Bofors in Sweden and Fiat in Italy.

From these leading industrial companies the dockyard purchased its machines and other advanced equipment for ships and workshops.

Engineers and craftsmen were travelling abroad on a strictly legal basis to work and to get training by these industries and they often made contributions to innovating products.

In the 1880s the Royal Naval Dockyard was among the first in Europe to introduce telephone supplies, electrical lighting and wireless telegraphy.

Conclusion and perspectives

Throughout the 18th century the Royal Naval Dockyard was the leading technological centre of the twin-monarchy.

Generally the period bears the stamp of crucial changes in the organizational structure. These innovative changes, having been described as an institutionalization and professionalization, together with the receptiveness to foreign technical innovations, account satisfactorily for the fact that, at the turn of the century, Danish warship construction was equal to that of any other country in Europe.

The many attempts made to put naval architecture into a scientific framework do not seem to have had any decisive influence in Denmark.

The establishment of the educational institutions made it possible to pick up or institutionalize groups inside the naval hierarchy, which had not yet found their position or permanent appointment. The fact that over a very short period these groups were organized into a corps became the cornerstone in the process of professionalization.

The presence of shipwrights seated among senior naval officers in the dockyard administration, and their ability to negotiate and win their case for innovative design testified to the social change. In my opinion these social and organizational developments are more important than the theoretical discussion among a very small group of learned individuals. In short the scientists did not become naval architects but the master shipwrights became "maritime engineers".

It seems to be a false strategy to look after one single cause to explain the superior quality of the Danish-Norwegian warships. It seems to be more fruitful to look at the matter as a combination of mutually dependent factors as described above, viz. the continuing modernization through professionalization, systematic collection and transfer of intelligence from abroad and – not to forget – the noble art of industrial espionage.

This paper has been a presentation of some aspects of the technological development in the twin-monarchy of Denmark-Norway in the 18th century. But what are the international perspectives and what can we learn from the cases presented?

Firstly, as already pointed out by professor Svante Lindqvist in the Swedish history of technology, technical development in the military industrial complex must be accentuated in the history of technology in general.[19]

Secondly, my research quite clearly shows that in the process of technology transfer during the 18th century industrial espionage played a major role, and that existing source material can provide us with significant information on the history of technology. I shall emphasize, as professor John R. Harris does in his paper, that an international widening of this kind of study could be of great value.

Finally, the paper tells us something about technological development and modernization in a small nation state with only a few mineral resources, that hardly produced any spectacular innovations and therefore was a country where technology transfer played a major role. This may also be true for other small European nation states and this calls for comparative studies.

Notes

1. A comprehensive bibliography on the history of the Danish-Norwegian Navy is found in Hans Chr. Bjerg: Dansk Marinehistorisk Bibliografi 1500-1975, Akademisk Forlag, København 1975. Unfortunately there is no literature on this subject in foreign languages. For a brief introduction see Hans Chr. Bjerg og John Erichsen: Danske orlogsskibe 1690-1860. Konstruktion og dekoration, København 1980, pp. 197-201 (English summary) and Ole L. Frantzen: Truslen fra Øst. Dansk flådepolitik 1769-1807, København 1980, pp. 141-146 (English summary).
2. For an excellent introduction to the navies in the Baltic during the 18th century see, Jan Glete: Sails and Oars. Warships and Navies in the Baltic during the 18th Century (1700-1815), in M. Acerra, J. Merino and J. Meyer (eds.): Les Marines de Guerre Européenes XVII-XVIIIe siècles, Paris-Sorbonne 1985, pp. 369-400.
3. A collection of constructional drawings, unique in its kind, is kept by the Danish National Record Office, Naval Archives. The collection consists of more than 15.000 items, covering the period 1690-1848. The collection includes a considerable number of French and English sheer draughts.
4. Frank Allan Rasmussen: Konstruktionsvirksomhed og skibbyggeri ved Søetaten 1692-1772. Unpublished dissertation, University of Copenhagen 1989.
5. R. Steen Steensen: Søofficersskolen gennem 250 år, 1701-1951, J.H. Schultz Forlag, København 1951, p. 86ff.
6. See Frederick B. Artz: The Development of Technical Education in France, 1500-1850, MIT Press, London 1966, p. 51 and 102, and M. Blanchard (eds.): Répertoire général des lois, décrets, ordonnances, règlements et instructions sur la marine (3. Vols.), Paris 1848-1859, and Jean Boudriot: Constructeurs et constructions navales à Rochefort aux XVIIe et XVIIIe Siécles, in: Neptunia No. 157, 1985.
7. The minutes of proceedings of the Commission are kept by the National Record Office, Naval Archives. Orlogsværftets Arkiv. Fol. reg. 154. Konstruktionskommissionens Forhandlingsprotokoller 1739-1862. The surviving part of the scale models, dating back to the 1680s, are kept by the Royal Danish Naval Museum in Copenhagen.
8. See Pierre Clément (eds.): Lettres, instructions et mémoires de Colbert publiés d'après les Ordres de l'Empereur, (8 Vols.), Paris 1864, and A. Anthiaume: Le Navire, sa construction en France..., Paris 1922.
9. In England an Academy were suggested in 1730. But the Academy established in Portsmouth never amounted to much./.../As a result, the Academy did not become important until the early nineteenth century; See Daniel A. Baugh: British Naval Administration in the Age of Walpole, Princeton University Press 1965, p. 99ff. A parallel to the Danish Construction Committee was established in England as late as 1832.
10. See John Fincham: A History of Naval Architecture, Scolar Press, London 1979 (reprint), pp. IX-LXXXIV.
11. The National Record Office, Naval Archives, Adm. 29, 1751-52, Adm 632, marts-juni 1755, Adm. 634, januar-marts 1765 and Adm. 634, august-december 1756.
    See also, Geheimeraad Friderich Michael Krabbes Levnets=Beskrivelse/.../skrevet af ham selv i Aaret 1791, København 1793, pp. 16-29.
12. Pierre Bouguer: Traité du Navire, de sa Construction et de ses Mouvements, Paris 1746 and H.-L. Duhamel du Monceau: Eléments d'architecture navale ou Traité practique de la construction des vaisseaux, Paris 1752.

13. See Daniel G. Harris: F.H. Chapman, The First Naval Architect and his Work, Conway Maritime Press, 1989, for further information on naval architecture in Sweden.
14. Fincham 1979, p. 156.
15. Based on primary material in the Danish National Archives and in the Birmingham Reference Library, The Boulton & Watt Collection. Birmingham.
16. For a more detailed account see, Flemming Steen Nielsen: Ildmaskinen på Gammelholm, in Helge Kragh (ed.): I røg og damp. Dampmaskinens indførelse i Danmark 1760-1840, Teknisk Forlag, København 1992, p. 35ff.
17. See Frank Allan Rasmussen: Fra Birmingham til København: Overførsel af Boulton & Watt teknologi ca. 1800-1810, in Kragh 1992, p. 54ff.
18. In 1817 a school for shipbuilders was established, and new technical disciplines were introduced in the curriculum.
19. See Jan Hult, Svante Lindqvist, Wilhelm Odelberg and Sven Rydberg: Svensk Teknikhistoria, Gidlunds Bokförlag, Hedemora 1989, p. 121ff.

The Critique of Industrial Technology in the Netherlands and other Western Countries in the Nineteenth Century

Dick van Lente

1. Introduction

As a first introduction to the cultural response of Dutch society to the coming of industrial technology, let me tell you a little anecdote. One night in 1842, a certain Frederik van Sorge – a grocer, radical journalist and municipal secretary in a small town in the Netherlands – lay awake worrying about the misery that machines and factories could bring about. While thinking about this, he had a vision. Five beings from outer space, having pity on mankind for the drudgery by which it had to make a living, descended upon the earth and installed a great many machines and robots, that took over all human labour. Within a short time this apparatus created a tremendous amount of very cheap products. But the aliens had forgotten one thing. Since the working people had lost their jobs, they had also lost their income. Therefore, they could not buy even the cheapest machine-made goods. Wealth concentrated in the hands of a small elite, while the rest of the people sank into poverty and had to rely on poor relief. Fortunately, the good beings recognized their mistake. They came back and destroyed all the machinery they had introduced. The people went back to work and once again earned their living.[1]

This little story suggests that both machinery and violence against its introduction were considered "alien" phenomena in the Netherlands, reflecting the fact that most machinery came from elsewhere – though not from outer space – and that violence occurred more in dreams than in reality. This does not mean that fundamental criticism of modern technology, including violence against machines, did not exist in the Netherlands, but that it was a rather isolated phenomenon.[2] In this paper I will discuss three critics of modern technology in the Netherlands and try to explain why their criticism remained so isolated. In doing this, it will be illuminating to compare their thinking and their position with that of critics elsewhere in Europe. I will therefore begin by giving a brief overview of the critique of modern, especially industrial, technology in western Europe.

2. Ideological reactions to industrial technology

The response to industrial technology has always been extremely divergent in both form and content. Both positive and negative reactions could be found in all strata of society. In this paper I will focus upon the critique, not the praise. Also, I am not interested in the whole gamut of critical ideas as such, but in *ideology*, that is, in the relationship between ideas and power, or in other words: the way ideas about technology structure the experience and actions of influential persons and groups. It is no accident that the emergence of the ideological movements which have shaped the politics of industrial societies – liberalism, conservatism, christian democracy, socialism and their various derivates – appeared shortly after the beginning of industrialization: they were interpretations of the sea-change that was taking place and they presented alternative marching routes into the industrial future.

Studies about the relationship between technology and ideology in the nineteenth century are too scarce to base a well-balanced overview on.[3] From what we now know, it appears that a fundamental shift occurred around the middle of the nineteenth century. Before that time, criticism of and resistance to modern technology was much more widespread than in the second half of the century. Before 1850 the reference point for critics was the pre-revolutionary society of the eighteenth century; thereafter the inevitable disappearance of the old order was more or less accepted and critics became more future-oriented. Let us now look at the first half of the nineteenth century. Criticism at this time came from mainly three groups: working people who were threatened by unemployment as a consequence of mechanization, conservative magistrates and politicians, and romantic writers.

The most impressive study of resistance to modern technology during the early industrialization that I know is Adrian Randall's *Before the Luddites*.[4] Randall describes how the introduction, from the 1770s on, of spinning jennies and scribbling engines met with fierce resistance from workers in some areas in the woollen goods producing districts of England. Often, troops were called in to protect the houses and machinery of the innovating clothiers against furious attacks of the crowd. These disorders led to a full-scale discussion about the social desirability of the new machinery. The workers hired lawyers to argue their case in the courts. They appealed to old statute law which prohibited innovations that would cause unemployment and that protected the apprenticeship system. They found support among some of the landowning elite, some of whom wrote pamphlets on their behalf. The innovating entrepreneurs then submitted the question to Parliament which, after much lobbying from both sides, repealed the old statutes in 1809. In this way, room was made for free industrial enterprise and innovation.

The case against machinery included several arguments. In the first place it was argued that people thrown out of work by machines would have to be supported by the community. This argument was shared by many landowners, who paid the highest poor rates. One of them wrote in a pamphlet: "A trade is valuable to a country in proportion to the number of hands it employs."[5] Since the new machinery would only line the pockets of a few clothiers while ruining the rest of the community, it should be forbidden. As to the apprenticeship system, the innovators rightly saw this as an impediment to the introduction of factories. But to the workers and the traditionalists it was an essential institution for preserving the social order, because it regulated the socialization of the young.

Against these arguments, the innovators pointed to the cotton industry, where the enormous output of cheap cloth had easily found an expanding international market. Therefore thousands of people had found work in textile factories and cheap clothes had become available to the working classes. To this optimistic picture of expanding markets was added the warning that if Britain failed to innovate, other countries would do so and would flood the market with their cheap goods. In their opinion, the welfare of the country left no alternative to the introduction of machines.

This pattern of arguments was repeated again and again in the first half of the nineteenth century. In England, conservative politicians such as Oastler and Sadler, representing the old landed aritocracy, took up the case of the handloom weavers and factory workers in the 1830s and 40s.[6] They too laid much emphasis on the destruction of the family by the factory system which would have disastrous effects upon the social and moral order. They pleaded for a tax on machinery and played a leading role in the Ten Hours Movement, which achieved a shortening of working hours in the 1840s.

These conservatives took some of their arguments from romantic writers such as Robert Southey, Thomas Carlyle and Mary Shelley, who explored the wider ramifications of the "machine age". Carlyle, for example, pointed to what he called the "mechanization" of all aspects of life in industrial society: the Lancaster system of basic education, modern Government, which put a premium upon efficiency instead of moral argument, and a materialistic, mechanical way of thinking, which was displacing the concern with what he called "Dynamics", "the primary, unmodified forces and energies of man, the mysterious springs of Love, and Fear, and Wonder, of Enthusiasm, Poetry, Religion, all of which have a truly vital and *infinite* character". "Men have grown mechanical in head and heart, as well as in hand", he wrote.[7] In *Frankenstein* (1818), Mary Shelley described minutely and hauntingly how the drive for intellectual and technical domination destroy love and life itself.

In other countries, for example in Switzerland and in the German states, violence against machinery also occurred, accompanied by similar debates. Here too, workers were supported by part of the elite. The German states are especially interesting, because already in the eighteenth century some of them actively supported innovating entrepreneurs, e.g. by granting them exemptions from guild restrictions. This policy was opposed by the landed aristocracy. Their most important theorist at the beginning of the nineteenth century was the romantic economist Adam Müller, who was a declared foe of modern technology. On the other hand, many government officials saw no alternative to industrialization if the country was to compete with Britain. But they also discussed ways of avoiding the social turmoil that industry had caused there. One of them was Franz von Baader (1765-1841), a mine official and glass manufacturer in Bavaria, who had observed the social impact of industry during a trip in England and Scotland between 1791 and 1796.[8]

Von Baader was a conservative thinker, in that he pleaded for a reinstatement of the old social order, with the clergy, aristocracy and bourgeois firmly in their own places. But unlike Adam Müller, he was in favour of technological development, which he thought was a necessary part of social development in general. Since every technological innovation brought with it social stresses, it should not, according to Baader, be left to private enterprise, but should be regulated by the state. He called for the establishment of a government agency for assessing the possible effects of innovations. Not only economic, but also social costs of an innovation, such as unemployment, should be considered when deciding whether or not to introduce an innovation. Possible victims should be indemnified. The state should also carry out large technological projects, such as roads and mining, which are useful for the whole nation. It should employ tarriffs to protect native industry. In general it should prevent that some economic interests would operate at the expense of others and ensure that the benefits of technological improvement would be evenly distributed.[9]

I remind you, that neither conservatives, nor romantic writers, nor craftsmen and workers were all opposed to technology (or to all machinery). The response was mixed in every group. In England, for example, most Tories remained aloof from the battle against industrialism. Outside England Romantic writers hardly paid any attention to technology and industry, which is not surprising, because in the heyday of Romanticism industrialization on a great scale only took place in

England. Some romantic writers wrote admiringly of industry and machinery. The German theologian Schleiermacher, for example, predicted that machinery would release men from drudgery so that they would be able to devote themselves to more spiritual pursuits.[10] Others were fascinated by the "sublime" spectacle of mines, foundries, and steam locomotives: the mixture of dread and admiration that the power, the fire and the smoke of machines evoked, became a fashionable theme.[11]

But what the examples I have given do show, I hope, is the importance of ideology. The introduction of machines has been resented and feared in many times and places. But these feelings could only develop into a political force of some consequence, to the extent that they were articulated in a kind of theory, that showed the wider social meaning of technological innovation and made this criticism relevant to wider strata of society than those immediately concerned. The strength of the English weavers' resistance derived from the coincidence of their interests with those of part of the aristocracy, but also from the fact that they were protecting a meaningful social order against destructive forces. And the example of Von Baader suggests how a critical attitude of Government officials could have led to something like what we now call technology assessment in the early industrial period.

During the second half of the nineteenth century industrialization and the waning of the old order were gradually accepted as inevitable. This reorientation can be explained by changes in the relationship between the new and the old elites, changes in attitudes among the industrial working class and the emergence of the modern labour movement. Both in Germany and in England rivalries between the landed aristocracy and the rising industrial bourgeoisie abated; there was a considerable mingling and cooperation between the two. In the industrial areas the number of workers who had never known anything but industrial labour grew. For them, the old social order was no longer the standard by which to measure social relations. Modern labour unions therefore accepted the industrial order and tried successfully to improve the position of the workers within that order, for example by claiming higher wages and shortening of the working day. Technology and the quality of work practically ceased to be a theme of discussion. Karl Marx, whose ideas were very influential in the social-democratic movement after 1880, argued that capitalism was a transitional phase, in which the means of production were perfected, creating tremendous wealth. Like the liberals, marxists considered the problems of industrialization as temporary symptoms of the transition to a better world. They directed their efforts towards the future, instead of lamenting the loss of old values. In this climate, critics of technology such as William Morris and Heinrich Riehl were considered marginal figures.

3. Three Dutch critics[12]

The first of the three Dutch critics whom I want to introduce to you is Frederik van Sorge (1803-1851), the radical journalist whom we already met at the beginning of this paper. In 1842 he published a pamphlet "about the influence of machinery upon the wealth of nations".[13] It was a contribution to a lively discussion about the problem of poverty, which was going on in several European countries. Although there were only very few really large factories in the Netherlands at the time, Van Sorge was worried about their introduction, for which academic economists were pleading. He warned that labour saving machinery would create mass unemployment, as it had done in England. This in turn would lead to the collapse of the market, even for cheap goods, which would happen even sooner when other countries would follow the lead of England and mechanize production. Widespread poverty would be the result.

Van Sorge was not against machinery in general. Mechanizing the process of production would enable the people to pursue higher, intellectual goals. Whether they would be able to earn a living in this way depended upon the market for books and periodicals, in other words on the level of civilization. Van Sorge therefore pleaded for the extension of education, which would create more opportunities for writers, journalists and other intellectual workers. Only when this change in the labour market would cause a shortage of manual labour has the time come for the introduction of labour saving machinery, he argued. Until that time the introduction of machinery should be impeded by means of a tax on machines. The returns of this tax should be used to carry out large public works, such as land reclamation projects and railroads, which would also create work for the unemployed. For each kind of work a minimum wage should be determined, based upon the needs of the worker. As long as international agreement upon these matters was lacking, Dutch industry should receive protection by means of tariffs.

Van Sorge's vision reminds us of Von Baader. Like the German official, he was not against machinery as such, but he rejected the idea that technological innovation should be left to private enterprise. The economy and technology should serve human needs and therefore had to be regulated. Van Sorge had only scorn for the argument that cheap products create a market for themselves, thereby enlarging employment: "But why, I ask, more consumption than necessary in order to live pleasantly? Is this not an abuse of the gifts of nature?"[14]

A critical attitude toward industrial technology was also prevalent in some government circles. In the early 1830s, for example, the government sponsored Dutch Trading Company set up a textile industry in the eastern part of the country (this was after the Belgian secession, when the Netherlands lost, in one stroke, their most important industrial area).[15] The secretary of the Trading Company, Willem de Clercq, and the English engineer who was to supervise the project, Thomas Ainsworth, agreed that they would not introduce power machinery, but the most advanced hand technology. Ainsworth wrote: "The neat cottage with its little garden and rosy-faced children in my opinion is a far more pleasurable sight, than to stand at a cotton mill door at 9 o'clock on a winter's night and contemplate the squalid looks of a few hundred poor wretches, who, whilst living, are dying."[16] The secretary of the Trading Company said that the Company aimed at a widespread cottage industry, which allowed workers to combine farm work with industrial labour. It must be added, that the cheapness of labour was another important argument for preferring hand technology.[17]

On the other hand, most Dutch academic economists were enthusiastic about modern industrial technology and urged Dutch entrepreneurs to follow the British example. All the well-known arguments of the political economists were echoed in the leading Dutch economic journals.[18] After 1870, a new generation of liberal intellectuals started to write prolific books about the social problems which accompanied industrialization. They argued that those problems should not be blamed upon machinery, but upon "human ignorance and immorality, obsolete institutions and laws".[19] Social legislation could, in their opinion, ensure that technology would serve the needs of all.

We need not go into the ideas of Dutch Socialists. Their opinions echoed those of Socialists in other countries. Whether radical or moderate, they all believed technical progress would benefit the workers, once they had assumed control of the state. Therefore, resisting technological progress was considered reactionary and workers were urged to direct their protests at improving their position within industrial society.

More typical of the Dutch situation is the reaction of the churches.[20] I will discuss two representatives of the most pronounced religious-political movements in the Netherlands, the Neo-Calvinists and the Roman Catholics.

The best example of a romantic critique of modern technology in the Netherlands came from the leader of the Orthodox Protestants (or Neo-Calvinists), Abraham Kuyper (1837-1920). Kuyper had been trained for the clergy at the university of Leiden, but he had turned away from the modernist theology which was being taught there and had experienced a conversion to radical Calvinism. He became the leader of the opposition of the Orthodox Protestants against the rule of liberalism in the state and the church. His campaign for government subsidies for denominational schools led to the first modern political party in the Netherlands (1879). When he could not convert the Dutch Protestant Church, he led the Orthodox Protestants in a secession (1886). He started his own newspaper and his own university, the still existing Free University of Amsterdam. I will first discuss a speech Kuyper held in 1869, in which he argued for the right of the orthodox protestants to their own place in society. This plea was embedded in a full-scale attack upon modern society, including modern technology, which often reminds us of Carlyle.[21]

The speech was titled "Uniformity, the curse of modern life". Modern society tends toward uniformity, Kuyper contended. Compare, for example, the endless variety of facades in the old towns of Holland, the charming irregularity of the streets and canals to the dead monotony of modern cities, with their straight streets and housing blocks. The modern state, another example, is ruled by centralized bureaucracies and a politician is no longer his own man, but has to conform to the party machine. Considering technology in a more material sense, the same tendency is visible: "A steam engine eliminates the rich variety which gave every trade something charming. The power of capital, concentrating in an alarming way, sucks away the life blood of our small enterprises."[22] Modern means of transportation and communication work in the same direction: they wipe out differences between peoples. This will result in "one big metropolis, which will know neither north nor south, east nor west; all life will be the same because it will show the uniform traits of death."[23]

During the 1870s and 1880s Kuyper slowly relinquished his extreme antimodernist position and especially his attitude towards modern technology. When for example he traveled through the United States in 1898, he reported enthusiastically to his newspaper about the skyscrapers, the electric streetcars, the comfort and speed of the trains and the telephone and typewriter in his hotel room.[24] When in 1896 the ore workers in the Rotterdam harbour struck against the introduction of electrical cranes, he called their action useless and unwise, because those cranes were part of a general process of mechanization which would eventually also benefit the workers.[25] In his theoretical writings after 1890, Kuyper claimed that it was a divine command that christians contribute to technical progress.[26] Like the Socialists and the Liberals of his time, Kuyper asserted that the problems often associated with technology should not be attributed to technology itself but to its uses, and therefore to the institutions and norms of society.

Neither Kuyper himself, nor students of his life and thought have explained this fundamental change in outlook. The most probable explanation is that Kuyper had changed his strategy. When he spoke about "Uniformity" in 1869, he was at the beginning of his career, taking the lead of the "small people", as they were called, the Orthodox Protestant craftsmen, labourers and small farmers, in their fight against the rule of liberalism. But Kuyper wanted much more than the vindication of the rights of this religious minority. He wanted to re-christianize the whole society. He realized that this could never be done by mobilizing the petty bourgeoisie against capitalism and socialism and by resisting modern technology and industry. Rather, he should use modern means for his own ends and try to win over capitalists and workers. He therefore incorporated some of the ideas of liberalism and socialism in his ideology; and among these was the wholesale acceptance of modern technology. This attitude has been well characterized by the theologian Haitjema: "The Neo-Calvinist wants to feel at home in the modern world; even more than that: he wants to master it."[27]

My third witness is A.H.J. Engels (1867-1940), a devout Roman Catholic weaver in Twente, in the eastern part of the Netherlands, where, as I just described, textile industry had been introduced during the 1830s. Engels was one of the founders of the first Roman Catholic trade union in 1889. In 1907 he published a very personal account of the advent of the factory system in Twente and in the same year he analysed the actions of the Rotterdam grain workers against the introduction of unloading machinery.[28] These publications show Engels' pessimism regarding the effects of modern industrial technology upon the workers.

The sketch "Fabrieksmenschen" (factory people) shows clearly that Ainsworth's and De Clercq's dreams about an idyllic textile producing population did not materialize. Life of the textile workers was characterized by poverty and exploitation. The introduction of the factory system after 1850 (when government interference in the economy was gradually abandoned) made things worse: the spinners and weavers, who had always supplemented their incomes with some farming, now became completely dependent on the factories and their harsh owners. Violent resistance against machinery, as described by Randall, did not occur here. This may be explained by the fact that employment in the textile industry rose much sharper than the population (thanks to the protected market of the Dutch colonies in the East Indies and a growing domestic market) and that, at least after 1866 for which period data are available, real wages also rose.[29] Nevertheless, there was an increasing number of strikes, sometimes accompanied by violence against the property of industrialists. These actions were often directed against the lowering of piece rates when more efficient machinery was introduced.[30]

Engels became convinced that technical development could not be stopped and that resistance against it made no sense. Later, when discussing the thousands of injuries and mutilations of working men, women and children reported by factory inspectors, he concluded that slowly but inexorably the army of invalids was growing, "victims of the industrial war of competition" in which only money counts. In his comments upon the dockworkers' strike against the introduction of unloading machinery in 1907, Engels suggested that municipal government should take charge of this, in order to make a more gradual transition possible.

But he did not sound very convinced that such a proposal would have any chance of acceptance. Other Catholic union leaders showed a similar pessimism.[31]

Engels was no original thinker, but his place in the Roman Catholic social movement is interesting. The ideas about technology of Roman Catholic leaders had evolved in a way very similar to Kuyper's. Before 1890, not many of them paid much attention to the problems of industry, but those who did were extremely critical. Following the lead of pope Pius IX, they rejected everything liberal and modern, such as the new secondary schools which had a lot of natural science in their curricula, and modern industry. But when Leo XIII became pope in 1878, the Roman church changed its strategy towards the modern world.[32] In his encyclicals *Aeterni patris* (1879) and *Rerum novarum* (1891) the pope proclaimed a more positive attitude towards science and technology and urged the clergy to help the workers create Roman Catholic labour unions that could stand up for both their material and their spiritual rights. From the 1890s on we see in Roman Catholic periodicals a very positive attitude toward modern science and technology. Technological innovation was presented, as it was by Kuyper, as a divine command. The direct links between technology and the misery of the workers, as documented by people like Engels, were denied: these problems could only be blamed upon the liberal-capitalist order of society, which misused the products of human intelligence, which were really a gift of God. Once society would be ordered along corporative lines, technology would benefit everyone.

Engels' writings show clearly that he was unable to reconcile this sanguine view with what he saw. But it is significant that neither he nor other union men ever openly challenged the official position of the church. Writing about the protests of the Rotterdam dock-workers against the introduction of unloading machinery in 1907, Engels said that technological innovation cannot be stopped, even though it costs thousands of workers their jobs; "and", he added, "for the sake of progress it should not be stopped". This apparent contradiction (if something cannot be done, the question whether it should be done becomes irrelevant) shows his unease with the official point of view.

4. Conclusion

At the end of the nineteenth century there was a remarkable convergence in attitudes and ideas about industrial technology among the main ideological currents in the Netherlands. Technological and social development were regarded as two separate phenomena: technology was the expanding arsenal of means by which man would improve his wealth and his mastery over the world. This would be to everyone's benefit, if only legislation would be improved to protect the workers, as the Liberals said, or if the workers would come to power, as the Socialists claimed, or if society was reconstructed along the lines of corporatism, as the Orthodox Protestants and Roman Catholics argued. The attitude of the Liberals and Socialists did not change significantly during the period studied, but the churches experienced a kind of conversion to technological optimism. I have tried to explain this as a strategic move, intended to strengthen their position in a secularising world, in which the onmarch of modern technology and modern labour relations seemed inevitable. This tendency may also be observed in other European countries, but unlike Germany and England for example, the Netherlands hardly had an undercurrent of criticism of modern technology. Why was this so?

I think part of the answer is to be found in the weakness of the Romantic and Conservative movements in the Netherlands as compared to surrounding countries. These movements were to a large extent a reaction to the Enlightenment, the industrial revolution and the French revolution. The Netherlands partook all these changes, but in a very moderate way, which explains at least partly why Conservatism and Romanticism did not strike deep roots in Dutch society. Since these movements produced the most radical and consistent critique of modern industry, their weakness in the Netherlands goes a long way to explain, why incidental protests against modern machinery did not find a sounding board in the discussions of the elites, who were, on the whole, convinced of the prosperity industry would bring the nation. In practice this meant that the shoemakers in Brabant who destroyed machines during the crisis of the 1880s and the dock-workers in Rotterdam who fought various innovations between 1880 and 1907 did not find the support of those who could have defended them in the courts or taken up their case in parliament, as happened in Britain and Germany. This is one of the reasons why the industrial revolution in the Netherlands, in spite of the great changes it wrought in society, did not provoke any sustained resistance.

Notes

1. F. van Sorge, *Over de invloed van machines op de welvaart der volken* (Middelburg 1842), 9-14
2. I. J. Brugmans, *De arbeidende klasse in Nederland in de negentiende eeuw* (Utrecht 1978) 183; J. van Meeuwen, *Zo rood als de roodste socialisten* (Amsterdam1981) 60.
3. The best studies have been done for England and the United States. See for England M. Berg, *The machinery question and the making of political economy* (Cambridge 1980); M. Berg, *The age of manufactures 1700-1820* (London 1985); A. Randall, *Before the Luddites* (Cambridge 1991); about the United States J. F. Kasson, *Civilizing the machine* (Harmondsworth 1977); L. Marx, *The machine in the garden* (London 1967). Germany: R.P. Sieferle, *Fortschrittsfeinde? Opposition gegen Technik und Industrie von der Romantik bis zur Gegenwart* (München 1984); Th. Hollenbach, *Lob und kritik der industriellen Revolution in England und Deutschland 1800-1848* (Frankfurt am Main 1990); W. König, "Die Rezeption von Technik und Industrie in der katholischen Romantik", in T. Pirker e.a. (Hg.), *Technik und industrielle Revolution* (Opladen 1987).
4. Randall, op. cit.
5. Randall, op. cit., 240
6. See for the following Berg, *Machinery question*, chapter 11.
7. Th. Carlyle, "Signs of the time" (1829) in *Selected writings (Harmondsworht 1980)* 67, 72.
8. König, op. cit.
9. Similar ideas, stressing the role of the state in regulating the economy, may be found in Fichtes *Der geschlossne Handelsstaat* (1800). See R. Kurz, *Der Kollaps der Modernisierung* (Frankfurt am Main 1991), 35-38.
10. König, op. cit. 247; Sieferle, op. cit. 51.
11. Kasson, op. cit. 166-168; F. D. Klingender, *Art and the industrial revolution* (1947) (Frogmore 1975), chapter 5; for the Netherlands see D. van Lente, J. W. Schot, "Heyenbrock en de Nederlandse industrialisatie" in W. Buitelaar e.a. (red.), *In het spoor van Heyenbrock* (n.p.n.d. 1988) 20-29.
12. See for more a extensive discussion of the Dutch response to modern technology my *Techniek en ideologie* (Groningen 1988) and "Ideology and technology. Reactions to modern technology in the Netherlands 1850-1920", in *European History Quarterly* 22/3 (July 1992), 383-414.
13. F. van Sorge, op. cit.
14. F. van Sorge, op. cit., 21
15. For the following, see R. T. Griffiths, *Industrial retardation in the Netherlands 1830-1850* (Den Haag 1979), 138-149; E. J. Fischer, *Fabriqeurs en fabrikanten* (Utrecht 1983), 61-90.
16. Griffiths, op. cit., 146.
17. Griffiths, ibid.
18. See for example the review of Van Sorge's pamphlet in the well-known journal *Tijdschrift voor staathuishoudkunde en statistiek* (1844) II, 129-140.
19. H. Goeman Bourgesius, "Stoommachines en volkswelvaart" in *Vragen des Tijds* (1876) I, 25.
20. It is hard to say exactly how typical, since hardly any research has been done about this subject in other countries.
21. If we understand technology as every kind of goal-directed organization, whether of matter or of human relations, the whole speech is really about technology. In this Kuyper resemples two other thinkers who have been strongly influenced by calvinism: Carlyle, who in "Signs of the times" spoke of the "mechanical age" and Jacques Ellul, who criticized "la technique".
22. idem, 19.
23. idem 21.

24. A. Kuyper, *Varia Americana* (Amsterdam 1899) 33-38.
25. *De Standaard* 25 March 1896.
26. See especially A. Kuyper, *De gemeene gratie* (Leiden 1902-1904) and A. Kuyper, *Pro rege* (Kampen 1911-1912)
27. Cited in S. J. Riderbos, *De theologishce cultuurbeschouwing van Abraham Kuyper* (Kampen 1947) 270-271.
28. A.H.J. Engels, *Fabrieksmenschen* (Leiden 1907); A.H.J. Engels, "De staking in de Rotterdamsche haven" *Katholiek sociaal weekblad* 26 oct. 1907, 492; "Gemeentelijke exploitatie van elevators" *Katholiek sociaal weekblad* 2 July 1910, 319.
29. Fischer, op.cit., 82, 164
30. H.D. Grobben, "Sociale conflicten en sociale organisaties in de Twentse textielindustrie 1860-1912" *Textielhistorische bijdragen* 12 (1971) 41-42.
31. E.g. H. Hermans, *Handboekje voor de RK werkliedenvereeningingen* (Amsterdam 1919 2nd ed.) 26, 27.
32. See B. McSweeney, *Roman catholicism. The research for relevance* (Oxford 1980) chapter 3.

Transatlantic Technology Transfer: The Reception and early Use of the Telephone in USA and Europe

Helge Kragh

Invention of telephony

The idea of transforming speech into electrical pulses so as to make it amenable to wire transmission goes back to the 1850s, when several people thought of how such a scheme could be realized. Charles Bourseul in France, among others, devised in 1854 a method for transmitting pitches of sound telegraphically, but it was only in the early 1860s that the first primitive apparatus of this type – a telephone – was constructed. It was done by Johan Phillipp Reis, a German schoolteacher, who invented an apparatus where the sound controlled the electrical current from a battery. Reis succeeded in transmitting single tones and, imperfectly, sounds over a narrow range of frequency; but the quality was so poor that there was no market for his invention, which was not further developed. In spite of Reis and other precursors, the merit of having invented a practical telephone belongs unquestionably to Alexander Graham Bell, the Scottish-American inventor and teacher of the deaf, who in 1874 envisaged an electromagnetic telephone where a membrane, actuated by the voice, produces electrical oscillations by vibrating in front of a magnet. In Bell's original scheme the transmitter and receiver were identical, and the electricity was supplied by the current induced by the permanent magnet in the telephone itself, which thus did not need an external battery.

Two years later Bell and his assistant, Thomas A. Watson, had developed his idea into an apparatus that enabled them to transmit complete sentences, the first one being the famous "Mr. Watson – come here, I want you." Bell received U.S. patent no. 174,465 for his invention of the telephone on March 7, 1876, an event which heralded a new age in the history of telecommunication. Several other inventors, including Thomas Edison, Elisha Gray and Amos E. Dolbear in America, Poul la Cour in Denmark and David Hughes in England, worked with ideas of electric telephones at the same time. Bell only filed his application for a patent a few hours before Gray, whose telephone used a separate transmitter, the electrical resistance of which was modulated by the voice.

The early pioneers of telephony all modelled their inventions after the telegraph, which was seen as *the* form of electrical communication. The mental confinement to the telegraph formed a powerful paradigm which deeply affected the entire development of telephony. This *telegraph paradigm* governed the inventive phase of telephone apparatus, the uses of the telephone, and also the engineers' conception of how telephone currents propagate in wires.[1] The mental identification of the telephone with the telegraph shaped Reis's conception of his invention and was also an important element in the telephone inventions of Gray, who, as a technical expert to Western Union, was deeply rooted in the telegraph community. Bell, too, was occupied with telegraphy, but his mental model of the telephone also drew on an alternative, acoustic tradition. Bell was, after all, a specialist in speech and had as much experience with phenomena of sound as with telegraphy. His position as an outsider to the telegraph community made him less bound by the telegraph paradigm and helped him develop an alternative vision based on his familiarity with speech rather than electrical pulses. As James Clerk Maxwell, the great physicist, pointed out in his Rede Lecture of 1878: "Prof. Graham Bell, the inventor of the telephone, is not an electrician who has found out how to make a tin plate speak, but a speaker who, to gain his private ends, has become an electrician."[2]

Bell's vision of an electric speaking tube notwithstanding, neither he nor his contemporaries could escape the telegraph paradigm completely. The telephone was generally seen as a species of the telegraph and in its early days often referred to as such. For example, when William Thomson, the later Lord Kelvin, first demonstrated Bell's apparatus in England in September 1876, he called it "the greatest by far of all the marvels of the electric telegraph."[3] The nomenclature revealed that the telephone had not yet obtained its own identity. As the apparatus was frequently called an "electric telegraph", so were telephone calls long referred to as "messages" or "dispatches".

In the spring of 1876 Bell had a telephone, but neither he nor others had very clear ideas of what to use it for. He offered Western Union the rights to exploit his invention for $100,000, but the big telegraph company rejected the offer because they were uncertain of its practical use in an era so thoroughly predominated by telegraphic communication. The telephone existed as an apparatus, an artefact, but there was no clear conception of what it actually was, i.e., for what purposes it should be used. To launch telephony as a means of speech communication required more than inventing apparatus, it also required a definition of the technology in the sense that its social purposes had to be invented. It was by no means clear that the telephone "is" a technology for general-purpose communication or that it was a conversational rather than, say, a broadcasting device. These functions had first to be perceived or invented until telephony, in the form of a telephone communication system, could come into existence.

The success of the Bell Telephone Company [4] relied to a large extent on Bell's and his associates' visions of telephony as a common communication service for ordinary citizens. Bell expressed his vision in a remarkable letter of March 25, 1878, to the "capitalists of the Electric Telephone Company" in London, a group of businessmen intending to exploit Bell's invention in England. In the letter he defined the telephone as "an electrical contrivance for reproducing, in distant places, the tones and articulations of a speaker's voice, so that conversation can be carried on by word of mouth between persons in different rooms, in different streets, or in different towns." He envisaged a telephone *system*, a network of telephones connected by wire or cable in the same way as houses were already connected by a network of water and gas pipes:

> It is conceivable that cables of telephone wires could be laid under ground, or suspended overhead, communicating by branch wires with private dwellings, counting houses, shops, manufactories, etc., uniting them through the main cable with a central office where the wire could be connected as desired, establishing direct communication between any two places in the city. ... I believe in the future wires will unite head offices of telephone companies in different cities, and a man in one part of the country may communicate by word of mouth with another in a distant place.[5]

If this does not seem much of a vision today, it is only because it has become reality a long time ago and we are now used to think of telephony as just the communication system Bell foresaw in 1878. At the time it was far from evident that Bell's vision was realistic or just desirable. But for the Bell Telephone Company it was a blueprint which guided the establishment of the first telephone systems in America and also helped create a different response to the telephone in America than in Europe.

Transfer to Europe

When the new invention was transferred to Europe, it was at first regarded with amazement, and more as a scientific curiosity than a practical technology. In scientific circles it was welcomed with enthusiasm, although more because of its potential use as a measuring instrument than because of its use as a communication device. Many scientists and electricians were disappointed to learn that the invention did neither rest on new scientific discoveries nor was mechanically advanced. An yet it worked so wonderfully.

When Thomson was first confronted with the telephone, as a judge at the International Centennial Exhibition in Philadelphia in July 1876, he was much impressed, but did not fail to point out that the marvel had been obtained "by appliances of a quite homespun and rudimentary character."[6] Maxwell recalled in 1878 that when he and his physicist colleagues had first heard of the telephone two years earlier "[we] began to exercise our imaginations in picturing some triumph of constructive skill." They had expected something at least as sophisticated as the siphon telegraph recorder. But, Maxwell went on,

> When at last this little instrument appeared, consisting, as it does, of parts, everyone of which is familiar to us, and capable of being put together by an amateur, the disappointment arising from its humble appearance was only partially relieved on finding that it was really able to talk.[7]

Whereas the Bell System concentrated on the practical development of the telephone and its incorporation into an extended communication system, scientific interest in the telephone was at first a European speciality. Leading physicists, including Hermann von Helmholtz, Ludwig Boltzmann and Max Wien, studied the theory of the telephone and the microphone, and the first exact theories of the propagation of telephone currents were worked out by Aimé Vaschy in France and Oliver Heaviside in England. Also the first uses of the telephone as a measuring device were due to European physicists. But the scientific understanding of how the telephone worked was not paralleled by an understanding of its social possibilities.

When looking at the technically primitive apparatus, telegraph engineers, used to advanced telegraphic machines such as the Wheatstone receiver and the Hughes printing telegraph, saw only a piece of toy with very limited practical possibilities; and certainly not a future threat to the almighty telegraph business. Looking back to the early days of experimental telephony, a British engineer recalled that "Within a few weeks of the receipt in this country of the first telephone instrument [I] had constructed a telephone with a couple of baking powder boxes and pieces of tin, and it had acted, for short distances at any rate, as well as the telephone of to-day."[8] Such experiences, shared by telephone enthusiasts all over Europe, did not enhance the social status of the new technology.

Western Union's refusal of purchasing Bell's patents in 1876 was probably beneficial for the development of telephony, for in this way American telephone business developed independently of telegraph interests and quickly became established as a communication service in its own right. The situation was different in Europe, where the telegraph services were owned and operated by government agencies of a centralized, bureaucratic and often conservative structure. Mentally and professionally, telegraph engineers and officials were committed to the telegraph which they saw as the one and only way to transmit messages reliably and efficiently over wire. In general, the post and telegraph administrations responded cooly to the telephone, which, in most cases, was seen as a curious and somewhat primitive telegraph of little interest as a public service. *The Times*, which earlier had characterized telephony as "American humbug", told its readers in 1877 that although the instrument might have a future in America, "In England ... it is hardly likely to become more than an electrical toy, or a drawing-room telegraph, or at most a kind of electrical speaking tube."9

This attitude was shared by many people, including telegraph engineers who had difficulties to imagine what real use the telephone could possibly serve. William Preece, who later became chief engineer of the British General Post Office, argued in 1879 that the popularity of the telephone was an American phenomenon depending on the particular social and cultural conditions prevailing in that country. In England, on the other hand,

> ... we have no difficulty in getting servants .... [but] the absence of servants has to a certain extent compelled Americans to adopt this system of telegraphy [telephony] for their own domestic purposes, and the telephone is to be found in almost every house as the only available substitute for the old system. Few have worked at the telephone much more than I have. I have one in my office, but more for show, as I do not use it because I do not want it. If I want to send a message to another room, I use a sounder or employ a boy to take it. 10

Preece was not, as one might believe from this quotation, a conservative telegraph engineer. On the contrary, in those early years of British telephony, Preece was known as a progressive defender of the new technology. He had brought the first pair of telephones to the country in 1877, and was recognized as an expert in the area. (Thomson had demonstrated a pair of Bell Telephones at the British Association for the Advancement of Science meeting in Glasgow in September 1876, but the instrument did not work properly). It would, however, be wrong to put all the blame for the slow European response to the telephone on the post and telegraph administrations. In England, as in most other European countries, the government did not consider the telephone just a toy. The Post Office realized the potential usefulness of the telephone earlier than most, but for a variety of reasons it was unable, or unwilling, to shape the conditions necessary for a quick development.11 Among the reasons was uncertainty about the telephone's relationship to the telegraph system, and the lack of a public demand for a national telephone service.

The restricted conception of the actual and potential usefulness of the telephone was not only a result of conservatism or a lack of imagination. It was also a result of the technical limitations of the early telephone. In 1880, when telephonic range was limited to 50 km at most, it was, after all, unrealistic to imagine a worldwide telephone network. The first generation of Bell telephones enabled people to speak only one at a time, and required silent rooms and good hearing. When Europeans encountered this imperfect apparatus, it is understandable that they associated it with a speaking tube of limited use. A typical example is provided by one of the first Danish descriptions of the telephone, given by Adolf Bauer in December 1877, shortly after the first experiments had been made in the country. Bauer was well informed and optimistic about the future of the new instrument, could it only be further developed. Yet his account of its use was narrow and lacking in vision:

> [The telephone] is useful in the same way as the speaking tube, where it finds application, but it is much simpler and cheaper; and it can be used over much longer distances than the speaking tube, where at present recourse has to be taken to [electrical] alarms or ordinary telegraph devices. The telephone offers a unique means of communication between various rooms within a house, from basement to upper floors, from warehouse to office, from one police or fire station to the other, etc.[12]

The mental identification of telephony with telegraphy, one aspect of the telegraph paradigm, turned out to be a major obstacle to Europe's development of telephony. In Great Britain, the attitude was legally confirmed in 1880, when the government charged the private United Telephone Company with infringing the Postmaster-General's monopoly on domestic telegraphy. According to the Telegraph Acts of 1863 and 1869 a 'telegraph' was "any apparatus for transmitting messages or other communications by means of electric signals," for which reasons Her Majesty's Government claimed that telephony was a species of telegraphy and hence the exclusive business of the Post Office.[13] The court decided that this was indeed the case. Although telephony was thus defined as a branch of telegraphy, the government wisely decided to issue licenses to private companies; but only under conditions which secured that the new technology would not threaten the telegraph system. For example, when the United Telephone Company acquired its license in 1881, its operations were limited by law to a radius of five miles.

In Europe, experiments with telephones were made by private enthusiasts and telegraph engineers in 1876-79, either based on imported Bell telephones or locally manufactured equipment. The initial transfer was not organized or planned, but took place in a rather haphazard way, relying on individuals who happened to be interested in the matter. In most cases, the telephone was brought to the European nations either by the way of travellers returning from America or via foreign newspapers and journals. This informal kind of technology transfer is what the Finnish historian of electrification Timo Myllyntaus has called "low-cost diffusion of easily accessible technology." It was possible to transfer telephone technology in this way because of the technical simplicity of the telephone and because Bell's original patents were restricted to the United States and easy to copy or develop by the recipients. For example, in Finland the telephone was pioneered by the engineer Daniel Waldén in 1877. All he needed was to read about Bell's invention in the Swiss *Journal Télégraphique*.[14] Similarly, in Germany the first Bell telephones arrived as a result of an article in *Scientific American* which caught the interest of Heinrich von Stephan, director of the Imperial Post and Telegraph Administration.[15] As mentioned, the first British telephones were gifts from Bell to Preece, who in 1877 was on a study tour to America, where he also met with Edison. Preece not only brought the telephone to England, but also supplied Stephan with the first two telephones. Shortly after Bell's invention, several European companies developed their own telephones, the most important being Siemens & Halske in Germany, who obtained their first telephone patent in 1877 after having successfully refuted Bell's complaint of patent right infringement.

The introduction of the telephone in Denmark may serve as another example of how telephones were first transferred to Europe.[16] The first telephones in the country were not, in this case, Bell instruments, but telephones bought in 1877 from Siemens & Halske. They were purchased by the State Telegraph for experiments, but after testing and demonstration for the Royal family the State Telegraph decided to leave the matter. Its director, Heinrich Høncke, deemed that the telephone was of no interest as a public service. Telephony was instead left to private enthusiasts, who also started a small local production of telephones. In 1879 a private telephone company was established in Copenhagen, which was purchased the following year by the newly founded Copenhagen Bell Telephone Company. In 1882 the company was transformed to the independent Copenhagen Telephone Company, which in the following years expanded to the nation's leading telephone business. In this development the State Telegraph showed no initiative or interest.

Swiss and Swedish experiences were about the same. The Swiss Telegraph Administration also obtained its first test telephones from Siemens & Halske in late 1877, and a privately owned company associated with the Bell System opened under government concession the first exchange in Zürich three years later. In Sweden, as in Denmark, the government Telegraph Board reacted passively after having tested the telephone in 1877. It left the exploitation to a number of local companies, some of them owned or controlled by the Bell System. However, contrary to Denmark, both the Swiss and Swedish telephone networks diffused very rapidly, and Sweden soon became Europe's telephone country number one. In the period of private telephone enterprises, the licensed companies paid royalties to the state except in the Scandinavian countries where no royalties were claimed.

In most countries the first telephone systems were established by small private companies. They operated under the constraints issued by the government, but the national telegraph offices preferred a passive role so long as the telephones did not cause trouble to the telegraph services. Long-distance telephony was sometimes assigned the post and telegraph offices, sometimes the private companies, but the structure and balancing differed from country to country.[17] Some of the first telephone companies, such as those in the Scandinavian countries, were opened as subsidiaries of the International Bell Telephone Company, and supplied with material from its European headquarter in Antwerp (the Bell Manufacturing Company, formed in 1882 by Bell and Western Electric). In most cases, they were soon taken over by local enterprises or the state.

A fairly typical example is France, where the telephone was first demonstrated at the Paris Academy of Science in October 1877. Models of more practical use were exhibited at the Paris exhibition in 1878, but without attracting much interest. The telegraph administration was unwilling to pioneer a national telephone system for which it could neither perceive strategic, political nor economic reasons. Instead the government granted concession to a number of private companies. After having consolidated into the Société Genérale des Téléphones in 1881, the new company created the first public telephone services in Paris with the government as an interested, but passive observer. As telephone business grew, the government became more involved, and in 1889 it took over the entire system. A roughly similar pattern was followed by other countries, so that most of European telephony was partially or completely nationalized at the outbreak of the First World War.

The introduction of the telephone in Germany differed from this pattern in that telephony was a state monopoly from the beginning. The first German lines were not urban, but suburban and rural, where they supplemented the existing telegraph system. Also contrary to other countries, the Telegraph Administration shared from an early date Bell's vision of a universal service accessible to ordinary citizens. The idea was promoted by Stephan in 1878, but not implemented. In this case the state telegraph bureaucracy was the visionary party, the popular attitude the hindrance. The prospective customers were simply not interested in joining a general telephone project at that early date, and flatly refused to participate in Stephan's dream.

What is a telephone?

As mentioned, Bell's telephone came on the market without any clear notion of what it could be used for. The applications and social attitudes attached to the apparatus had first to be established until the technology was properly defined, and people knew what a telephone was.

The telephone, understood as an instrument of conversation, only emerged gradually and with different pace in America and Europe. The first lines were either private lines, with two offices or homes in direct connection, or lines connecting the stock exchange, opera or telegraph offices with telephone booths. Only later, between 1879 and 1883, were central exchanges introduced, which made possible a wider conversation use of the telephone. But even then it took time to conceive the telephone as a natural means of communication. Bell's vision of the telephone included its use as a social technology, enabling people to socialize at a distance, but such use was rare in the early years of telephony and only became common in America about 1895. In Europe it took much longer to adapt etiquette and social customs to the telephone, which until the 1920s continued to be regarded a somewhat inappropriate instrument for real conversation. Only people of the lower classes would extend or accept a dinner invitation by telephone; and, as a rule, those people did not own a telephone.

The difference in the social and cultural attitudes to telephony between the New and the Old World was described vividly in 1912 by Arnold Bennett, a British telephone expert and author. "What strikes and frightens the backward European almost as much as anything in the United States is the efficiency and fearful universality of the telephone," he wrote, characterizing the United States as "primarily nothing but a vast congeries of telephone cabins." He continued his comparison as follows:

> I do not mean that Europe has failed to adopt the telephone ... but I do mean that the European telephone is a toy, and a somewhat clumsy one, compared to the seriousness of the American telephone. Many otherwise highly civilized Europeans are as timid in addressing a telephone as they would be in addressing a royal sovereign. The average European middle-class householder still speaks of his telephone, if he has one, in the same falsely casual tone as the corresponding American is liable to speak about his motor car.[18]

There were several reasons for the European backwardness, a theme constantly appearing in the popular and technical literature of the period. Perhaps the most important one was the telegraph paradigm, which ruled stronger and longer in Europe than in the United States. Initially, the telephone was sometimes conceived as simply a telegraph accessory, a means for transmitting speech between telegraphers. Many of the first telephones were connected to the telegraph stations so that customers could phone their messages instead of sending a messenger or going to the station themselves. In Germany, as in England, the telephone was officially adopted as a type of telegraph system.

In the 1880s the telegraph was not only a superior long-distance communication system, but also, it seemed to many, entirely satisfactory for communication over short distances. The telephone allowed instant reply, but then the telegraph left a written record for reference and was able to send more words per minute than even the fastest speaker could pronounce. In England and some other countries telegraphy had developed into a variety of forms, including systems of local intercommunication between customers. With the "ABC telegraph" people could communicate by ordinary Roman letters typed on a keyboard without caring about the Morse code, and other systems ("pantographs" or "pantelegraphs") transmitted handwriting telegraphically. Other telegraphs again were specially constructed for fire alarm, stock price information, crime protection or the latest news. The excellent service offered by the existing communication system, the national post and telegraph services, was a main reason why Europeans found it difficult to appreciate the usefulness of the telephone and for a long time considered it a luxury. Furthermore, the state-owned telegraph companies were often reluctant to engage wholeheartedly in telephony because they feared that progress in the new technology would take place at the expense of telegraphy.

This was particularly important in long-distance communication, an area where the telegraph used to reign supreme. The Belgian engineer François Van Rysselberghe's system of composite telephony and telegraphy over long distances was rejected in England because it was seen as conflicting with telegraph interests. Van Rysselberghe obtained his results by modifying the telegraph signals, which facilitated the transmission of telephone currents but slowed down the speed of the telegraph signals. Preece was "unable to understand the advantage of any gain in speaking on a wire which is detrimental to telegraphic communication," and said that "we cannot afford in England to cripple the one system in order to benefit the other."[19] As it were, this attitude did cripple one system, telephony, to the benefit of another, telegraphy.

Eventually, the telephone did develop to become a cheap and democratic means of communication, available to almost everyone, but this feature of universality was not there from the beginning. Early telephony was expensive and largely limited to the business community and wealthy families. In 1896, the fee for service in New York City was $20 a month, which was about half the average income of a worker. At that time the Bell System started a successful campaign to democratize the telephone and turn it into a universal service system, a mass instead of a luxury product. However, even in progressive America true universality was seen as an unrealistic goal and telephony still regarded a means to satisfy the needs of the middle class and the business community. In 1905, an optimistic Bell estimate assumed that saturation would be reached at 20 telephones per 100 Americans.[20]

Telephone subscription was as expensive in Europe as in America, and also in Europe the service was mostly used by businessmen. In 1881, the first year of Danish telephony, a subscriber outside central Copenhagen was charged an annual fee of 200 kroner. In comparison, the salary of a female telegraph operator was about 1000 kroner a year.

In many European countries, the social attitude to the telephone continued to be that it was a luxury technology, a privilege of the rich, and not a practical necessity. In America, "by the middle '80's, a city physician who did not have a telephone lost much business and was apt to be regarded as a back number."[21] As late as 1892 only six out of Copenhagen's 458 physicians had found it appropriate to install a telephone in their office. Arnold Morley, the British Postmaster-General, argued in 1895 that telephones were not and never would be necessities for the average citizen in the same way as were gas and water. The telephone-as-a-luxury attitude was widespread and accepted. According to *The Times*,

> The telephone is not an affair of the million. It is a convenience for the well-to-do and a trade appliance for persons who can very well afford to pay for it. For people who use it constantly it is an immense economy, ... [but] for those who use it merely to save themselves trouble or add to the diversions of life it is a luxury. An overwhelming majority of the population do not use it and are not likely to use it at all, except perhaps to the extent of an occasional message from a public station.[22]

At that time subscribers in London were charged £17 per year for unlimited service, only slightly less than it would cost to employ a maid.

Educating the public

When the first public telephone exchange was established in Berlin in early 1881, headed by the industrialist Emil Rathenau and promoted by Stephan, the telegraph director, the initial number of subscribers were – eight. It took half a year of hard work to persuade another 88 prominent citizens to join the project. And this was in a big city, a bustling centre of capitalist activity. In the more remote areas of Europe the outcome of the efforts was more modest. When G. Gilmore came to the Isle of Man in 1881, he was eager to introduce the telephone. "I ... tried to get the natives to sign agreement to take telephones, but could not get a single one to do so, and at the finish I had to set up a number of free lines to educate the people into the thing."[23] In other words, there was no public need for the telephone.

One notable difference between American and European attitudes to the telephone was the way in which the telephone companies perceived the needs of their customers. In America, the strategy of the Bell System was aggressive in that the company realized that if telephony should expand, it was them, and not the customers, who should set the agenda. Whereas the American telephone policy was supply- rather than demand-oriented, European telegraph administrations and private telephone companies held a much more passive attitude, seeing their roles in responding to people's needs rather than creating them. In America, the middle class was eager to obtain a telephone, and was encouraged to become subscribers. In Europe it was rather discouraged.

Creating a market for the telephone was, at any rate, an urgent problem for the young telephone industry. There had never been a public demand for telephones and when the technology came on the market, it was not clear what it would be used for. People therefore had to be carefully educated so as to appreciate the technology and, eventually, see it as a necessity. Such education or propaganda took place both in America and Europe (as well as elsewhere), but with great difference in intensity, persistence and planning. From an early date the Bell System made heavy use of advertising, information campaigns, public demonstrations and other kinds of "education". Compared to the American campaigns to create needs, European efforts were modest, scattered and incoherent.

The theme of needs and education, and Europe's inability to copy the aggressive American entrepreneurial spirit, continued to be a central question in the development of European telephony. In 1869 a British engineer acknowledged "that our American brethren – perhaps on account of their admitted spirit of enterprise, and go-aheadness – are much more appreciative than we are of the benefits to be obtained from the telephone."[24] The gulf between Europe and the United States in this area only widened in the following years. Forty-three years later, in 1922, another Briton, Frank Gill, chief-engineer of the Bell-owned International Western Electric and President of the British Institution of Electrical Engineering, summarized the situation as follows:

> Is communication and even better communication a necessity or not? If it is, then the passive attitude of merely satisfying public demands must be abandoned and an aggressive attitude take its place. ... In the United States, in

particular, there has for years been an educational campaign of a very high order, coupled with the construction of plant in advance, without which it is, of course, worse than useless to create demand. All this seems self-evident and trite, but two questions will test very quickly whether in fact this matter is quite so self-evident and obvious as it appears; these questions are – I. Has telephony, during the 46 years it has been available, been of as much use to Europe as it might have been? II. Have the organizations, Government or otherwise, been permitted to do what they have wished to do? The answer to both questions is most decidedly – No. [25]

Telephony as pleasure: Theatrophones and electrophones

The earliest experiments with telephones, such as Reis's and Bell's "musical telegraphs", were associated with the transmission of tones and music rather than speech. For some time the future of the telephone was seen in its capacity of an instrument broadcasting music or news. Throughout the formative period where telephony's true nature as a universal two-way communication and conversation system was established, it continued to be used as a broadcasting technology, that is, one-way mass communication. Although it was used as such on both sides of the Atlantic, this use was more common in Europe and can to some extent be seen as reflecting social and cultural attitudes characteristic of the Old World.

Shortly after Bell's invention reached Europe, the first experiments with telephonic concerts took place. In December 1877 the Technical University in Vienna experimented with transmission of songs and music over a distance corresponding to 30 km. Bell telephones, without separate microphones, were used, which must have resulted in poor quality of transmission. But newspapers, reporting that "every note could be appreciated, and the tone was phoned with extraordinary purity," were enthusiastic. At the 1881 Paris electrical exhibition, the electric light was the big hit and Edison the celebrated figure of the new electrical age. The telephone did not attract much interest as a speaking apparatus, but big crowds queued up before the rooms where entertainment was transmitted live from the Opera and the Théâtre Français. The Munich electrical exhibition of 1882 offered long-distance transmission of music from Oberammergau, located 97 km from the city. In subsequent technical exhibitions such telephonic performances were standard components.[26]

In the early 1880s, European telephony had not yet stabilized as a system of practical communication. The very concept of telephony was in an unsettled state, which caused some concern as to the future use of the telephone. An editorial in

The Electrician described in 1882 the state of British telephony as "petty", being "little more than an improved system of speaking tubes capable of offering great facilities for personal communication between various points in one locality, but not capable, as yet, of competing with the telegraph despatch lines."[27] The journal was worried that this state of affair might not change and the telephone then degenerate into a toy. It referred to

> ... the project of tapping the stage and the pulpit telephonically, or of laying on operatic music, like gas, for the use of every householder. There may be people in this world who go to operas and theatres and churches only to listen; and for such as these the fact that you can actually hear an orchestra or a voice, or a voice and a tune, at one and the same time if only you have faith, and nobody bothers you, may be exceedingly interesting, perhaps even profitable; but that a Government department would enter upon this kind of business is, to say the least of it, problematical.

Entertainment arrangements, where the telephone was used as a medium of multiple address for music and theatre, were frequent all over Europe, and often served as a first entrance to the new technology. For example, one of the first telephone lines in Copenhagen connected the Chamber of Industry with the Royal Theatre, thereby providing the capitalists the opportunity of listening to the opera during breaks in their transactions. Other early lines connected restaurants with music halls and orchestras. In Copenhagen's fancy Café Elektrique telephones were placed at the tables, not for table-to-table talks, but for the reception of music. In Paris, a system of so-called theatrophones were in operation in the 1890s. The Theatrophone Company offered subscribers connection to Paris theatres, and also set up a number of coinboxes where the public could listen to five minutes of theatre performance for one franc.

The telephone served similar functions in England, where the affluent citizen could listen to various London entertainments over the telephone. And affluent they had to be, for in 1896 the cost for installation and one year's service was £15. Although music was the most popular subject for such telephone broadcasting, some services also offered news, church services, recital of poems and sport. The Electrophone Company in London, which was established in 1894, offered a variety of entertainment, but no news service. Another kind of telephonic entertainment was offered by the National Telephone Company, which in 1892 connected several of the Glasgow churches with their local exchange. In this way, "their subscribers whose houses are connected with the exchange could hear the services by telephone, and asking permission to place a telephone instrument or instruments in St. George's Church."[28]

By far the most sophisticated and successful of the European theatrophonic services was the Hungarian Telefon Hirmondó, which was created in 1893 by Tivadar Puskás, a former Edison employee, and which continued until the early 1920s. The Hirmondó telephone broadcasting system had at its height about 6000 subscribers in Budapest. Unlike the theatrophone and electrophone, it offered a wide spectrum of information and concentrated on news rather than entertainment. It operated very much like a newspaper and prefigured later electronic media such as radio and television. Political news, stock exchange reports, railway information, weather reports, sport and local news were brought seven days a week in alteration with concerts and other musical performances. It was even possible to have a phonograph attached to the telephone receiver, and in this way record the programmes for later use. In spite of its popularity, Telefon Hirmondó was in no way a service intended for the average Hungarian, and exhibited much of the same class structure which characterized other forms of European telephony. The audience was the Budapest aristocracy and economic elite, which was reflected in the priority of the programmes. The largest share was devoted to detailed stock exchange news, both within the Austro-Hungarian monarchy and from foreign exchanges. Hirmondó was an authorized service expressing the views of the ruling elite, and it required the permission of the government to become a subscriber.

Why was the world's only successful telephone newspaper Hungarian, whereas corresponding attempts in America and elsewhere in Europe either failed or were short-lived? The early general European interest in broadcasting telephony reflected an uncertainty about the proper use of the telephone. When European nations gradually developed their telephone systems along the American line, that is, towards an extensive system of conversation, interest in the pleasure function waned. In the highly bureaucratic and authoritarian Hungary ordinary telephony remained undeveloped, and interest limited to a small upper class with plenty of money and leisure time. The pleasure or news telephone system fitted better to a static class society than a dynamic, democratic society. That, at least, was the explanation given by a British observer in 1898. In Budapest, a city of pleasure and easy leisure to the fortunate few, people could afford the luxury of the Telefon Hirmondó. But in London, "a man could not sit the whole day with the apparatus to his ear waiting for some particular news or exchange prices."[29] In New York, he could even less.

The differences in American and European attitudes to telephony manifested themselves in the ways the telephone was used. Generally, American telephony expanded rapidly, both quantitatively and with regard to new areas of use. Interurban and long-distance lines opened at an early date, and before 1900 the telephone had become a middle class communication instrument. In Europe, telephony was for a long time restricted to intra- and interurban use, long-distance lines were fewer and shorter, and the middle class only adopted the telephone after World War I. As to new uses of telephony, three areas may be singled out. None of these were particularly important, but if compared with the one European speciality, the pleasure phone, they illustrate the drive and inventiveness of American telephony.

In the big American cities, the hotels quickly became centres of electrical communication. Telephones in every room enabled guests both to call other rooms and hotel services, and also gave access to outside lines. The concentration of telephones in the hotels amazed European visitors, who often described the bustling skyscrapers, with their numerous elevators and electrical lines, as one big, fearful efficient organism. In his above-mentioned report, Bennett wrote: "Just as I think of the big cities as agglomerations pierced everywhere by elevator-shafts full of movement, so I think of them as being threaded under pavements and over roofs and between floors and ceilings and between walls, by millions upon millions of live filaments that unite all the privacies of the organism – and destroy them in order to make one immense publicity."[30] And there was good reason to be impressed. By 1904, the Waldorf-Astoria in New York could boast of 1120 telephones and 500.000 calls a year; the hundred largest hotels in the city had more telephones than Spain.[31] European hotel industry stuck to messengers in order to satisfy the requests of their guests. In big hotels this was a complicated procedure which required many messengers and employees. But in Europe there was no scarcity of labour, and most guests valued the personal contact more than electronic efficiency.

The first extensive use of electrical technology in the political process came in connection with the American presidential elections of the 1890s, when election returns were gathered and coordinated by means of the telephone and telegraph networks. This was first done in 1892, and in the 1896 election the Bell System organized a national telephone election service which resulted in a quick and efficient return. There was no such large-scale political use of telephony in Europe.

Military applications of telephony was a vision of many generals at the turn of the century, but a widespread introduction of telephones in the armed forces met resistance among officers who feared that the more informal nature of telephone conversation would undermine their authority. The first practical use of telephony took place during the Spanish-American war, when the bombardment of Santiago de Cuba by US warships was directed by telephoning from the front to the shore, from where the range and direction were flagged to the fleet. The commander of the Army Signal Corps, general Augustus Greely, had the entire front interlaced by telephone lines and submarine telegraph cables laid between Cuba and the United States. In this way he was able to communicate quickly with

Washington and better coordinate the military and political strategies. Later, during the First World War, American telephone technology proved its military value and superiority to the European attempts to exploit telephony militarily.[32]

Telephone culture

As mentioned, telephony was not intended for the average citizen or for casual conversations. The telephone companies primarily thought of it as a serious instrument of business and encouraged functions which paralleled those traditionally belonging to the telegraph. They were never happy about the entertainment telephone arrangements, which were mostly operated by small, independent firms. The repression of social calls may be seen as another result of the telegraph heritage. Such "calls" did not match with the technology of the telegraph, and were virtually unknown in telegraphy. Consequently social conversations tended to be considered unfit also for telephony. However, subscribers did not always follow this view, witness the popularity of theatrophones and the like.

As the telephone industry and administrations saw it, many people lacked the proper respect for the new instruments, which they "abused" for less worthy purposes such as chats and exchange of gossip. Already in 1877 Bell had suggested the socializing function of the telephone, which he associated with female subscribers in particular. He predicted a time when Mrs. Smith would spend an hour with Mrs. Brown "very enjoyably in cutting up Mrs. Robinson" over the telephone.[33] But such use was ill-regarded both by the telephone companies and the serious (male) business subscribers. It was only in the 1920s that telephone industry, first in America, began to endorse sociability and emphasize that the telephone was, after all, more than just a practical technology.[34]

Women occupied a special position in early telephone culture and tended to be seen as abusers rather than users. Not only were their conversations, as everyone knows, talkative, redundant and frivolous. They were also ignorant about the technical aspects of telephony and the proper way to handle the instrument. In the early days of telephony, companies as well as customers were preoccupied with how to use the telephone correctly, which seems to have been a more important subject than what to use it for. The handling of the apparatus and, in particular, the correct way of speaking were important and much discussed elements of the telephone culture. Journals encouraged their readers to speak quietly and dignified (don't shout!), directly into the microphone, and to use the telephone as an instrument of improving speech articulation and good manners. "We have observed … that those who use the telephone have very much bettered their articulation and enunciation of words," wrote an American electrical journal i 1884.[35] In the same year a Danish journal expressed concern about the manners and oral skills of the girls who operated the newly established exchange in Copenhagen. Complaints about the operators' lack of elocution were common, but the journal believed in the educational effect of the exchange: "The telephone service has an ennobling effect on the speech organs of the young ladies; it helps them to develop a distinct pronunciation, to include all the consonants of the words, and to articulate clearly the vocals."[36]

Not only was the women's mode of communication – invariably described as cheerful, chatting and unserious – discouraged by the telephone authorities, the transmission of female voices also seemed to be resisted by the very nature of speech current propagated through a wire. With the technology of the nineteenth century, voices were inevitably distorted and attenuated by the wire or cable, and the effect was the more marked the higher the pitch. In an instruction of how to use the telephone, the director of the Copenhagen Telephone Company, Christian Madsen, concluded in 1888 that "high, thin voices should not be allowed to telephone at all." He did not mention female speakers explicitly, but advised customers to "speak with a deep, but full and articulate voice directed right at the transmitter," an advice most women would have trouble living up to.[37]

The telephone culture involved many more elements than the few mentioned here. How was morality and social order influenced by the telephone? Was it possible to defend one's privacy and social distance against disreputable persons? Was it socially acceptable to use the telephone for this or that purpose? Wouldn't the popularity of the telephone threaten the established class order of society? Where should the telephone be placed in the home and what kind of apparatus was the most appropriate? These and other questions were eagerly discussed in the period, both in America and in Europe and essentially in the same way. The only notable difference seems to be that the preoccupation with etiquette and social distance declined in America many years earlier than in Europe.

Cultural determinants in technology transfer

The telephone entered the world as an apparatus without a definite use or plan of development. In Bell's bold vision it was conceived as an instrument of general communication, and in its subsequent development by the Bell System this goal was gradually realized. In its American context, it was embodied with expectations which reflected cultural values specific to the American zeitgeist.

When transferred to Europe, to a continent with different cultural values, the telephone appeared as a different technology. It was conceived differently, used for different purposes, and developed much more slowly. The striking technology gap between the New and the Old World in this area was not a result of the telephone being an American invention; patent protections were of little importance, and the time-lag, about one year, too little to account for the differences. Neither can the different paces of development be explained in terms of superior American technical knowledge. The telephone was a technically simple apparatus, and Europe was as well, if not better, equipped with competent scientists and telegraph experts. Organizational differences, especially the lack of unitary and independent European telephone companies, were of some importance, but not decisive. The main reason lay in the different zeitgeists, the attitudes and expectations to the telephone, that is, in cultural factors. That the cultural ambience codetermines technological development is generally accepted and not, of course, particular for early telephony. Technology is not culturally neutral, and during transfer an array of cultural factors may inhibit (or, conversely, enhance) the development of a technology.[38] In his analysis of the British backwardness in electrification, Tom Hughes described the difference between America and Great Britain in terms of zeitgeists, and contrasted the British "spirit of times" with the American "go-aheadness".[39] I have argued that a similar explanatory scheme – vague as it is – may be adopted in explaining the transfer of telephony.

One way of conceptualizing and generalizing the transfer and early development of telephone technology is to conceive a certain technology as a binary system consisting of what I shall call a "material core" and a "cultural belt". The core is made up of the artefact as such, that is, a collection of material objects connected in a certain way.[40] It is "surrounded by" a belt or sphere of social, economic and cultural ambients which determine the non-material context in which the technology functions. The cultural belt refers, first and foremost, to the actual, intended or potential uses of the technology, but also comprises an array of mental factors, attitudes and expectations. Although I depict this function as a belt, it is not external or grafted upon the technology. It is part and parcel of it. If the belt changes, the technology changes as well.

In general, there is no one-to-one relationship between the two components of a technology. An artefact, or core, is initially, at its invention and immediate implementation, associated with certain attitudes and expectations of use. These intended uses are, for example, spelled out in the patent application. But as the technology is subjected to the conditions of the market, the uses will often differ from those originally envisaged. The belt changes. In this sense, a technology may change substantially even if materially it remains unchanged.

An emerging technology has often an unsettled character in the sense that nobody knows what it "really is", because it is not known what it can be used for or what it is wanted for. This is typical for the many inventions that are not made as responses to definite needs or for definite purposes. Although the core may exist in a technically perfect form and is well understood, the invention is not yet defined. It is there, but without a clear identity. In this phase the technology *obtains* its identity by a gradual fixation of its belt. Of course, its core will develop too, be improved in perpetual interaction with changes of the belt. The settlement or stabilization of the belt is a relatively open-ended process in that it is not predetermined by the core. It is not completely open-ended, though, for it is subject to the technical limitations and possibilities given objectively by the constitution of the core. A horseshoe may be used for many things, and cultural attitudes to it may change in many ways, but whatever the attitude it cannot be used as, say, an aeroplane. What I want to emphasize with this model or metaphor, is, first, that technology is not a fixed thing with a fixed purpose; and, second, the importance of visions and cultural attitudes in the very definition of technologies.

A somewhat similar conclusion has been drawn by Claude Fischer in his perceptive study of the diffusion of telephony in America. The importance, and acceptability, of residential calls for social purposes is now taken for granted, but was ignored for almost half a century. In Europe it took even longer. Fischer's study "suggests that promoters of a technology do not necessarily know or determine its final uses; ... [and that] in promoting a technology, vendors are constrained not only by its technical and economic attributes but also by an interpretation of its uses shaped by its and their own histories, a cultural constraint that can be enduring and powerful."[41]

When a technology, which has not yet reached its settled stage, is transferred from one society to another, the core remains at first unchanged. But very often the belt will change under the pressure of the new cultural environment, so that the transferred technology differs from the original one. One way of expressing this cultural dependency of technology is to depict transfer as being "filtered" through a sociopolitical and cultural contextual net. This is, for example, an important feature in the model suggested by Myllyntaus.[42] What happens during the filtering process, we may now say, is a change in the cultural belt. But whereas Myllyntaus' model suggests that a transferred technology is merely used differently in different sociocultural contexts, I would like to suggest that the sociocultural filter often changes the technology itself. In its transitory stage, the technology is defined by its cultural belt. The "filter" model suggests a preexisting, stable cultural environment which externally acts upon the technology; by allowing only certain technologies, or modifications thereof, to pass the filter, the environment shapes the uses and organization of technologies. The same connotation is associated with the term "cultural constraint" as used by Fischer and others. I would rather express the relationship as an interactive one in which the cultural environment manifests itself in new forms which are attached to the technology and become parts of it. A new technology does not merely encounter a stable cultural environment to which it has to adapt itself. It stirs up the environment and creates new forms of culture related to the technology and its use.

Notes

1. For an analysis of the telegraph paradigm, and technological paradigms in general, see H. Kragh, "Om paradigmer i teknologien og udviklingen af teknologisk viden," *Polhem 9* (1991), 249-77.
2. J. C. Maxwell, "The telephone," *Nature 18* (1878), 159-63; reprinted in W. D. Niven, ed., *The Scientific Papers of James Clerk Maxwell* (New York: Dover Publ., 1965), vol. 2, pp. 742-55, on p. 751. For detailed analyses of the invention of the telephone and the role of the telegraph paradigm, see M. E. Gorman and W. B. Carlson, "Interpreting invention as a cognitive process: The case of Alexander Graham Bell, Thomas Edison, and the telephone," *Science, Technology, & Human Values 15* (1900), 131-64, and D. A. Hounshell, "Elisha Gray and the telephone: On the disadvantage of being an expert," *Technology and Culture 16* (1975), 133-61.
3. Quoted in F. G. C. Baldwin, *The History of the Telephone in the United Kingdom* (London: Chapman & Hall, 1925), p. 14.
4. Bell Telephone Company merged in 1879 into National Bell Telephone Company, which the following year was changed to American Bell Telephone Company. In 1885, American Telegraph & Telephone Company (AT&T) was incorporated as American Bell's subsidiary and long-distance branch. The two companies consolidated in 1900 under the name the Bell System. In this paper, I refer to the Bell System without distinguishing between the various phases and companies of its organizational history.
5. The letter is reproduced in its entirety in J. K. Kingsbury, *The Telephone and Telephone Exchanges* (New York: Longmans, Green, and Co., 1915), pp. 89-92.
6. Quoted in J. H. Robertson, *The Story of the Telephone* (London: Pitman & Sons, 1947), p. 8.
7. Maxwell, *Scientific Papers*, p. 752.
8. S. Evershed in *The Electrician 59* (1907), p. 273.
9. *The Times*, August 21, 1877.
10. In discussion of A. Scott, "Recent improvements in Professor Bell's telephone," *Journal of the Society of Telegraph Engineers 8* (1879), 327-45, on p. 337.
11. C. R. Perry, "The British experience 1876-1912: The impact of the telephone during the years of delay," pp. 69-96 in Ithiel de Sola Pool, ed., *The Social Impact of the Telephone* (Cambridge, Mass.: MIT Press, 1981).
12. A. Bauer, "Telefonen," *Illustreret Tidende 19* (1877), 92-94.
13. Baldwin, *History of the Telephone in the United Kingdom*, pp. 43-51.
14. T. Myllyntaus, "Transfer of electrical technology to Finland, 1870-1930," *Technology and Culture 32* (1991), 293-317. "La téléphonie," *Journal Télégraphique 3* (September 25, 1877), 668-69.
15. E. E. Feyerabend, *50 Jahre Fernsprecher in Deutschland 1877-1927* (Berlin: Reichpostministerium, 1927), p. 24. *Scientific American 37* (October 6, 1877).

16. For information about early Danish telephony, see M. Gredsted, *Statstelegrafen 1854-1904* (Copenhagen: Statstelegrafen, 1904) and V. Jarløv, ed., *Københavns Telefon i 75 År* (Copenhagen: KTAS, 1956).
17. For data on the ownership and early history of telephone systems in European countries, see Arnold R. Bennett, *The Telephone Systems of the Continent of Europe* (London: Longmans, Green, and Co., 1895). A summary account is given in F. Stumpers, "The History, development, and future of telecommunication in Europe", *IEEE Communications Magazine* 22:5 (1984), 85-95. Useful surveys of the development of telephony in the United States, France, and Germany are given in R. Mayntz and T. P. Hughes, eds., *The Development of Large Technical Systems* (Frankfurt a. M.: Campus Verlag, 1988).
18. *Harper's Monthly 125*, July 1912, p. 192. Here quoted from Sola Pool, ed., *Social Impact of the Telephone*, pp. 152-53.
19. W. Preece, "Recent progress in telephony," *Nature* 25 (1882), 516-19, on p. 519.
20. C. S. Fischer, "Touch someone: The telephone industry discovers sociability", *Technology and Culture* 29 (1988), 32-61, on p. 57.
21. Alvin F. Harlow, *Old Wires and New Waves: The History of the Telegraph, Telephone, and Wireless* (New York: Appleton-Century, 1936), p. 394.
22. *The Times*, January 14, 1902, as quoted in Perry, "The British experience," p. 75.
23. Quoted from Baldwin, *History of the Telephone in the United Kingdom*, p. 110. For the first Berlin telephone company, see Feyerabend, *50 Jahre Fernsprecher in Deutschland*, p. 29. See also J. Hoppe, "Vom Spielzeug zum Netz: Szenen aus der Geschichte des Telefons," *Kultur & Technik* 3 (1990), 39-45.
24. Scott, "Recent improvements," p. 328
25. F. Gill, "The future of long-distance telephony in Europe," *Electrical Communication* 1:2 (1922), 8-26, on p. 9.
26. Ernst Ruhmer, *Neuere Elektrophysikalische Erscheinungen* (Berlin: F. & M. Harwitz, 1907), pp. 161-73. For the use of telephony for entertainment purposes, see Carolyn Marvin, *When Old Technologies Were New: Thinking about Electric Communication in the Late Nineteenth Century* (New York: Oxford University Press, 1988), and Asa Briggs, "The pleasure telephone: A chapter in prehistory of the media," pp. 40-65 in Sola Pool, ed., *Social Impact of the Telephone*.
27. "The telephone and the state," *The Electrician* 8 (1882), 184-85.
28. *The Electrician* 28 (1892), p. 372.
29. Briggs, "The pleasure telephone," p. 53.
30. Quoted in Sola Pool, ed., *Social Impact of the Telephone*, p. 152.
31. S. H. Aronson, "Bell's electrical toy: What's the use? The sociology of early telephone usage", pp. 15-39 in Sola Pool, ed., *Social Impact of the Telephone*, on p. 30.
32. See H. Kragh, "Telephone technology and its interaction with science and the military, ca. 1900-1930," in Paul Forman and José M. Sánchez-Ron, eds., *National Military Establishments and the Advancement of Science and Technology: Studies in Twentieth Century History* (Boston: Kluwer; forthcoming).
33. Quoted in Aronson, "Bell's electrical toy," p. 31.
34. Fischer, "'Touch someone'".
35. Quoted from Marvin, *When Old Technologies were New*, p. 90.
36. *Illustreret Tidende*, January 27, 1884.
37. Christian L. Madsen, *Om Telefon-Ligningen: En undersøgelse af Telefonvæsenets Udvikling* (Copenhagen: Høst & Søn, 1888), p. 21. German summary in *Elektrotechnische Zeitschrift* 9 (1888), 462-68.

38. For a general discussion, see John M. Staudenmaier, *Technology's Storytellers: Reweaving the Human Fabric* (Cambridge, Mass.: MIT Press, 1985), pp. 121-61.
39. T. P. Hughes, "British electrical industry lag, 1882-1888," *Technology and Culture 3* (1962), 27-44.
40. The materiality is not essential, though. I use the term "material core" merely to emphasize the permanence and physical objectivity of the core.
41. Fischer, "'Touch someone'," p. 61. See also the much expanded treatment in C. S. Fischer, *America Calling: A Social History of the Telephone to 1940* (Berkeley: University of California Press, 1992).
42. Myllyntaus, "Transfer of electrical technology".

Early Rural Electrification in Denmark – a Reaction from People outside the Town Establishment

Hans Hedal

1. Introduction

Modern industrial society is characterized by the emergence of a division of labour between those who change and those who use technology. This is contrary to the pre-industrial society, where the change and use of technology was not divided. This division involves a risk of a disharmony between the functions of changing and using technology. Early electrification in Denmark was an example of this kind of disharmony. In the last decades of the 19th century early Danish electrification was limited to towns and big factories. Private investors, electrotechnical expertise and political authorities in fact electrified in a way, which did not include greater parts of industry and definitely not the rural population. Farmers in the cooperative movement and above all the related technical experts found this way of electrification unsatisfactory, because it did not include the rural areas.

Planning, innovation and spread of new technologies are not necessarily in harmony with the wishes of the users of technology (producers and consumers). Three developments of modern high technology, all related to its strongly increased complexity, have pushed this problem to its extreme. Firstly the constantly increasing time lag between the start and the realization of a project. Secondly the immensely increased amount of ressources involved. Thirdly the increased distance and isolation between planners, investors, technicians, scientists and politicians on the one side and ordinary citizens on the other. The first group of people may for a long period have "baked on" a mega-project in high technology, which at its realization does not get the necessary sympathy from the public. The planning of nuclear power and the mistrust among ordinary consumers of electricity is an example of this kind of high-technologial disharmony.

The possible disharmony between the functions of changing and using technology may provoke a reaction from the users of technology. That under certain favourable circumstances is a success. This was the case during early electrification in Denmark. Farmers and related expertise did not want electricity to benefit the towns and factories only. They reacted by making their electrification independent of town and factory electrification. Technological renewal-activities among farmers and related expertise, first and foremost Poul la Cour, were essential for early rural electrification in Denmark. In the period 1891-1902 la Cour and his assistants performed a process of innovation. And in the following decades: the 1902-1920s, small local DC power stations were spread into the country-side. They were based on wind, water and steam power and above all internal combustion engines, especially the diesel engine. This innovation and spread of electrical technology independent of electricity supplied from towns and industries constituted early rural electrification. It prepared the way for further rural electrification based on large scale high voltage AC technology.

In this article electrification is considered as 1) a transition from one form of energy supply to electricity, 2) an extension of an electric network, and 3) an extension of the supply and consumption of electricity. The fundamental aim of the article is to explain how early rural electrification took place in Denmark. The question can be dissolved into three main questions, namely:

1) What was the direction of the existing way of electrification and what caused it?
2) What were the electrification projects of the cooperative movement among farmers and related expertise?
3) To what extent did they succeed in implementing their projects and what conditions permitted this?

These questions are worth considering for at least three different reasons:

I) The study of early Danish rural electrification is interesting because it is atypical. Important for and specific to Danish electrification is the relatively high frequency of processes, where the involved popular movements and their related expertise were pursuing goals opposing the existing way of electrification, and to some extent they succeeded in realizing these goals. Danish history of technology contains three significant cases of such atypical developments. The first one is the above mentioned early rural electrification between the 1890s and the 1920s. There are indications that Denmark was among the earliest of nations in this respect. This and the use of wind-electricity at the turn of the century combined with the involvement of a popular movement and its related experts makes the development atypical. The second case starting in the 1950s is the plan to introduce nuclear power. The history of Danish nuclear power is atypical because the movement against nuclear power, related expertise and public opinion succeeded in avoiding this technology unlike the situation in the most other industrialized nations. The third case is the activities of the wind-energy movement and industry in the 1970s and the 80s which is atypical too: The alternative energy movement and related expertise succeeded in making Denmark a world leader in development, use and production of electricity-producing windmills. In short these three developments are the deviations which together make up the peculiarities of Danish electrification.

II) Moreover this important step in the development of modern Denmark is still relatively unexplored. Up till now the written presentations have either treated different sides of the subject [1] or have focused on la Cour's life.[2] A complete picture is still misssing. The purpose of this article is

1) to be part of the background material for a Danish history of technology, which is in course of preparation,
2) to summarize and unify existing knowledge in the field (the references in footnote 1 and 2),
3) to put early Danish rural electrification in its social, cultural, political and economic context,
4) to provide important additional material
5) to identify important unanswered questions in need of further research.

III) The study of early rural electrification in Denmark is interesting from a technology policy perspective. The study of this and the two other cases mentioned above may throw light on two important questions relevant for technology policy in modern industrial society: 1) Under which historical conditions will a disharmony emerge between the functions of changing and using technology – a disharmony that may provoke a reaction from the users of technology? This question is politically interesting if you want to make the development and use of modern technology more harmonious. 2) If popular movements and their related expertise find the existing line of development of a certain kind of technology undesirable, what can they do to change this line of development, given a certain set of historical conditions? This question is politically interesting, both if you want to hinder and promote such technology strategies.

2. The existing way of electrification 1879-1910

The Danish state was not at any time during this period involved in production and distribution of electricity. In fact, this did not occur before the 1920s where the government, quite exceptionally, subsidized the electrification of Southern Jutland. The Danish Government regarded the running of power stations as a private enterprise. Planning of power stations was left to commercial and municipal initiatives. As a result, only electricity supplies that appeared commercially profitable were established.

The DC power stations of that period based on steam or gas engines were relatively big enterprises. AC technology and diesel engines were not satisfactorily developed before the beginning of the 20th century. Moreover, these stations were only profitable in town areas or in big factories, where the consumption was big enough. In town areas the consumers were sufficient in number and situated sufficiently close to the power station. Because the low voltage resulted in a big power loss along the lines, DC power stations only had a radius of delivery at about 1 km. In rural areas a sufficient number of consumers were not available. For that reason the necessary profitability could not be obtained.

The first Danish industrial company which introduced electric lighting in 1879 was the machine factory and shipyard Burmeister & Wain. In the 1880s and in the beginning of the 1890s many private lighting systems were installed, many of them on larger factories.[3] It was the rapidly growing provincial towns which at the end of the nineteenth century introduced centralized power stations. A joint-stock company started the first proper production of electricity in Odense in 1891 with a 300 kW DC power station. Shortly afterwards a 840 kW power station was established in Copenhagen. The stations were intended to work as power suppliers to nearby electric lighting. During the 1890s a number of similar works were established in five provincial towns: Ålborg, Vejle, Hjørring, Nakskov and Kolding. The public power stations during the 1890s were mainly private enterprises, working with a municipal permission, but there was no economic relationship between municipality and company. Electrification was carried out by private investors and electrotechnical expertise. Also it was basically considered luxury consumption without any significance for population and production in general. Electric motors still had a marginal significance in Danish industry and craft at the end of the 1890s. In Copenhagen planning of new power stations continued, the second was completed in 1897 and the third in 1902. Århus got its first power station in 1901 and during the next twenty years all provincial towns joined. Skagen as one of the latest in 1918. In 1905 there were 43 public power stations in total, however with a low degree of distribution to consumers. In the beginning of the 20th century a new development was under way: Municipal takeover of power stations. This was normally taking place, when the power station wished to expand. Often there was not enough funding for such an investment. Consequently the municipalities would frequently enter. Gradually they saw it as a municipal task and as a source of income.

The towns did not start rural electrification and no initiatives at all came from the Danish government to start it. The only thing that happened in the field of rural electrification was not a result of a government initiative. It was the result of the initiative of the farmers and the outstanding effort of the physicist, inventor and teacher Poul la Cour at Askov Folk High School in Jutland. It was his iniative as an inventor and applicant for government subsidies which awakened the interest of the government during the 1890s. Only then the government began a modest subsidizing of activities preparing rural electrification, which were almost exclusively performed by la Cour.

During the first decade of the 20th century high voltage AC technology was satisfactorily developed. In Denmark AC was introduced for the first time in 1907 in Copenhagen and a suburb, Skovshoved. In the years 1910-1914 farmers began to get interested in AC power for their farms and established several cooperative companies possessing their own power supply. The provincial towns, which for quite a number of years had had their own DC power stations, very soon realized the advantages of supplying high AC voltage to the surrounding rural districts. High voltage AC production equipment was installed in urban power stations and high voltage electricity was delivered to transformer substations in surrounding districts. Now for the first time town municipalities, town investors and electrotechnical expertise electrified in a way which included at least some minor part of the rural population.

In the period 1879-1910 Danish political authorities, private town investors, town municipal authorities and electrotechnical expertise therefore only electrified in a way which certainly did not include the rural population.

3. Poul la Cour's rural electrification project [4]

Poul la Cour's means and ends 1891-1903

Poul la Cour realized the above-mentioned limitations of the existing way of electrification. But he also knew the immense potentialities of electrotechnology. He wished to promote a complete rural electrification in Denmark and in that way improve rural economy. The migration from country to towns should be stopped or reversed by making country and towns equal in use of electricity. As a solution to the problems he would try to develop a small and inexpensive electric power supply for the benefit of the rural population. Denmark had not any significant indigenous energy ressources except maybe wind. Poul la Cour imagined that the wind could be converted to electricity by a windmill coupled to a dynamo. He wished to store wind power, so it could be used when required for lighting, heating and motive power. As early as 1891 he expressed this wish in a letter to his brother. He also wished to contribute to the development of small industrial applications of electricity for the rural areas. At Askov Folk High School la Cour helped by his assisting co-workers made a series of technical solutions to problems of

1) transmission and transformation from mechanical to electric power
2) storing of electrical power
3) applicability of electric power
4) optimizing the production of electric power from rotor wings

The innovative work with wind power was susidized by the government. In the end la Cour's persistence made great results. He was the Danish scientist before 1910, who recieved by far the largest government subsidies. According to R. Skovmand the reason was that politics changed in the mid 90s from conservative rule to a negotiation policy lead by moderate forces of the leftist peasant party and the conservative right. The peasant party was in favour of the folk high schools. The conservatives wanted to support academic science.[5] The contents of government policy relating to la Cour's activities, its social power base and its significance at the time are subjects in need of further research.

Working with problems of storing, transport and applicability

In 1891 la Cour put forward an innovative idea about storage of wind energy, to make it available when needed for lighting, heating and motive power. Electricity should be produced as mentioned above by a dynamo driven by a windmill. Until then la Cour and others had worked on storing electrical power in storage batteries. However, this solution was too expensive, the durability was too poor and wind power was too unstable to work as a supply for such batteries. The idea was to use electric power to decompose water in oxygen and hydrogen, store the hydrogen in a gas engine and use the burning of hydrogen for illumination, heating and motive power when required. Parliament subsidized this invention with a modest sum of money. The same year a minor experimental mill was built. The mill was staffed by the two physics teachers at Askov Folk High School: la Cour and his assistant at the mill, Jacob Appel. A minor wooden building containing a millroom and a lab was erected around the 11 meters high mill tower.

Here research and development work began. The idea of storing hydrogen was postponed, until another limitation of windmills had been solved. Storage of wind power depended on the ability to smooth out and stabilize the rotation of the wings. For that reason la Cour started inventing a device, which converted the unstable movement of the wings to a steady and stable motive power. He made purely experience-based experiments with a simple mechanical arrangement of weights, pulleys and gear wheels. In 1892 he had developed a mechanism for smoothing and stabilization. This device was called a "Kratostate". Poul la Cour took out a patent on the kratostate, but renounced his right for kratostates constructed for regulation of wind, water and horse power in favour of the Danish people. Later la Cour refined this invention and made it more simple. He also developed an electromagnetic switch disconnecting the dynamo and the storage battery in case there was no wind.

The government subsidies continued in 1893, but on fresh conditions. Now money should be spent on experiments concerning the applications of wind power. Then la Cour began utilizing an Italian method of hydrolysis with the aim of using hydrogen as a medium for energy storage. In this way the problem of energy storage was solved easily and cheaply. In 1895 it was possible to produce oxyhydrogen gas for illuminating the rooms of Askov Folk High School. The windmill pulled a dynamo which was electrically coupled to a hydrolysis apparatus. A kerosene engine was used as a reserve. The produced oxyhydrogen gas was stored in 2 gas containers connected to a gas pipe distribution system. This type of system was cheaper than a system using storage batteries. Oxyhydrogen gas however was not as versatile as electricity for energy supply. For example, oxyhydrogen gas could not be used in electric motors; and la Cour like many others expected great things from electric motors. The most obvious solution to this problem was converting the gas back to electricity, most simply, perhaps, via an internal combustion engine. In 1895 he abandoned such experiments.

He now concentrated on making his second electrochemical process. As a result la Cour became the first in Denmark to produce calcium carbide. The method had been invented by the American T.L. Wilson and was very cheap. The only significant cost was electricity. Strong current had to be transmitted through a mixture of pulverized lime and coal. An inflammable gas, acetylene, was made by adding water to the produced calcium carbide. Here was a promising process which might contribute to the solution of the problems of storage, transport and applicability of wind power. At the same time it might contribute to the development of small industrial applications of electricity for the rural areas. Later in 1895 experiments performed by la Cour had shown, that the profitability of the process was doubtful unless a very strong electric current was at hand.

Other attempts at creating small industrial electrochemical technologies for rural areas were made. Poul la Cour wished to use a number of new technologies like welding with oxyhydrogen gas, production of sodium hydroxide and chemical fertilizer. However none of these attempts turned out successfully.

Optimizing the production of electric energy from the rotor wings

From 1896 to 1900 la Cour tried to optimize the production of electric energy from the rotor wings. The practical problem was how to optimize the shape, the number and the size of the wings of a windmill. In 1896 la Cour made a series of experiments with small scale models of windmills. Inspired by ongoing experiments in Copenhagen la Cour made a small wind tunnel and used his kratostate to make a steady air flow. He concluded that the open windscreens were preferable, because a minor extension of the length of the wings would soon make a 4-winged mill just as efficient as a 16-winged mill, and 12 wings would be saved.

In 1896 la Cour applied for further government subsidies for extension of the experiments in Askov concerning the applications of wind power. He wanted to make the smaller windscreens more efficient and develop various small industrial electrochemical technologies for rural areas, thus making bigger windmills competitive with the urban steam mills. Poul la Cour asked for grants for the construction of a bigger experimental mill including laboratory and workshop. In addition he asked for funds to employ an assistant and a miller and to expand his own activities as experimental leader. In the still favourable political atmosphere he recieved an additional grant and a considerable increase of the annual sum covering running costs.

The new bigger experimental mill was not finished until the turn of 1897-98. The mill was well-equipped and luxurious. A very big storage battery was placed in the cellar. The electricity was produced by a 2.5 kW and a 5.5 kW dynamo driven by the mill. Part of the electricity was used for hydrolysis.

The practical problem: how to optimize the shape, the number and the size of the wings of a windmill, was not only a technical one. It involved a series of more fundamental scientific questions in aerodynamics. The qualitative results of the experiments with small scale models of windmills were obvious, but the quantitative results were very uncertain. Poul la Cour viewed these experiments as purely introductory. In a series of more fundamental scientific experiments made in 1899 la Cour had determined the resultants of pressure on plates of different shape and inclination relative to the direction of the wind. From these he calculated the yield from different existing and imagined windmills. These experiments confirmed traditional experiences of millwrights. Broken profiles were the best and implied great technical possibilities. From these and the earlier experiments in 1896 la Cour concluded that the 4-winged mill was the best for practical purposes and that the breadth of the wings should be between 1/4 and 1/5 of their length. His experiments had shown that a model of an optimal mill could produce 4 times as much power as contemporary farming mills. Just as important, la Cour described a method to calculate the yield of other existing or imagined milltypes. The first mill constructed by means of these optimizing principles was the above-mentioned new big experimental mill, which was finished at the turn of 1897-98.

Final development of an alternative electrical technology

At the turn of the century the innovative activities of la Cour were criticised by the Danish electrotechnical expertise in their newly established journal. He was accused of being something like a technological romanticist. In 1901 the suggested small power stations were criticised for being uneconomical, and his research and development for being a waste of money. Danish social-liberals and their newspaper (Politiken) complained that government subsidies had been granted to la Cour. In 1902 the critique resulted in diminishing subsidies.

So la Cour was forced to prove that the experimental mill was useful for Danish agriculture. In 1902 la Cour converted the experimental mill to a power station for the village of Askov. The capacity of the experimental mill was increased. The number of storage batteries was doubled. Electricity was produced by two dynamos, each 6 kW. One of these worked as a reserve capacity driven by a kerosene engine. Later in 1902 la Cour applied for a government grant to carry out the final development of his alternative electrical systems technology, proposing small power stations as the one at Askov to satisfy the needs of villages. The many small windmills at the farms he proposed should be converted to small power stations at approximately 1.4 kW including a storage battery. This system included transmission through an electric grid on the farm, supplying electric lighting, motors and other appliances. One big motor driving the threshing machine, and a small portable one, which could be used at various places for various purposes, such as driving a chaff cutter, a grinding mill, a beet slicer, a band saw or a pump. These rural power systems were small and decentralized compared with contemporary urban power systems. The transfer of energy was very flexible compared with older forms of energy supply. This was the essential advantage of using electricity. Subsidies were then approved for another year, but without automatic prolongation. At the end of 1902 la Cour was fully convinced, that his wind power systems technology was sufficiently developed to be put into extensive use in villages and farms.

Spreading the knowledge of electricity 1902-1919

In advancing the spread of small DC power stations la Cour followed several courses of action from 1902 until his death in 1908. If people in a village or a provincial town planned a power plant, la Cour was often invited to explain the pros and cons of electricity. These meetings resulted in a lively correspondence between la Cour and interested people. His books about wind-electricity were widely read. The most important means of promoting rural electrification was the Danish Wind Electricity Society: DVES, which he founded in 1903. DVES published a bimonthly journal: "Tidsskrift for Vindelektrisitet" on wind electricity, la Cour being the author of most articles. Originally in 1903 the aim of the society was to offer guidance in construction of electric power stations essentially based on wind power; moreover lectures, publications and courses were offered. The members were electrotechnical experts, generally interested citizens and above all farmers – high and low. These people actively participated in promoting the spread of small rural power stations. Poul la Cour and DVES knew that their campaign would ride on a wave already sweeping the country-side. Since the beginning of the 1880s the cooperative movement were spread all over rural Denmark. Rural power stations would fit nicely into the cooperative pattern. During the years of its existence the aim of DVES changed concurrently with the changes in challenges. In 1905 the aim was broadened to offer guidance in construction of electric power stations in general. The objects clause was broadened again in 1907. Now the aim was to offer guidance in construction of electric power stations in general and in reconstruction of older plants. In 1912 DVES withdrew from projecting of power stations and concentrated on advising and spreading the knowledge of electricity. This was confirmed in the objects clause from 1914. DVES was finally dissolved in 1916. The reason was the development of the internal combustion engine, strongly limiting the advantages of wind power.[6]

Another means to promote the spread of small rural DC power stations was the training of rural electricians. At that time it was only in a few cities that one could be trained as an electrician. In 1904 DVES offered an educational programme for rural electricians. DVES instructed nearly 20 electricians a year at Askov. They learnt theory for three months and the maintenance and development of the Askov wind power station as parallel practical training. They then finished their education by building a small power-plant somewhere in Jutland. This was a short training period compared to the 4 years of the city electrician. An investigation of their later carreers shows that most of them found jobs as rural electricians. Many became self-employed and some owners or managers of power stations. The education was finally closed down in 1919.

La Cour versus the Danish electrotechnical expertise

In the last months of 1905 the leading Danish periodical of electrotechnology was dominated by the discussion about small power stations. As earlier mentioned Danish political authorities, private town investors, town municipal authorities and electrotechnical expertise from 1879 to 1910 only electrified in a way, which did not include the rural population. As might be expected the majority of Danish electroengineers were worried about the unplanned spread of these small plants in the country-side, finding them unprofessional and uneconomical. And they were worried about the training of rural electricians, mistrusting their technical competence. But they acknowledged small rural power stations as a necessary temporary measure, until the rational solution of high voltage AC could be supplied all over the country. Leading electrotechnical experts however could not see the advantage of using windmills for motive power. They found windmills producing electricity especially uneconomical. Poul la Cour answered the critique. He found farmers and rural electricians sufficiently technically competent. During the last quarter of the 19th century the cooperative movement had demonstrated the abilities of the farmers. In addition he presented a more optimistic view of the economy of small electrical plants. That quite a few electrical technicians supported the idea of small wind power stations, is suggested by the great number of engineers joining DVES.

In 1904 a commission concerning power plants was appointed by the Danish government. Here the above mentioned discussion about the technical competence of rural electricians was on the agenda several times. Poul la Cour was a member of the commission and fought strongly to maintain the right of local councils to authorize electricians. A subcommittee of the commission proposed that this right was taken away from local authorities and replaced by a state authorization combined with a examination. In the final law of 1906 a compromise was made. The right of the local authorities to approve electricians was maintained. But the installation of especially dangerous electrical equipment was made by electricians, holding a state authorization. This compromise was not only required to back up DVES interests represented by la Cour, but an unconditional requirement of state authorization would sincerely have delayed rural electrification.

A more serious problem for la Cour and DVES was the opposition from the Danish Union of Electricians, canalized through the Social Democrats. At the parliamentiary reading of the budget for 1907-1908 the Social Democrats criticised that no visible results had been achieved by la Cour's experiments. They also criticised the rural electricians for not being as technically competent as town electricians. The Social Democrats proposed that la Cour should recieve only a one year appropriation. But the proposal could not get a majority vote. He got a five year appropriation, which however ended too early, due to la Cour's death in 1908.

In 1914 DVES started negotiations about authorization of rural electricians in the rural districts. A new educational programme for rural electricians was proposed. The course should take two years instead of three months, now including all relevant knowledge about AC. The question was to be decided in the commission of electricity, which was appointed by the government as a result of the strong current-law in 1907. The commission would only approve the new education, if the course was concluded by a test similar to the test for town electricians. This demand was not accepted. The plans were shelved in 1919. At the same time training of rural electricians stopped at the experimental mill.

4. The farmers' electrification project

Before electrification the majority of farms employed far more manpower than today. The use of horses was also important. Other significant types of energy consumption was heating of rooms and cooking of food. Very common was burning of wood and peat, mainly from the farmers' own sources. A form of energy consumption that could not be satisfied by ones' own production was lighting. Before the advent of electricity lighting was obtained from tallow candles or kerosene, which in rural Denmark were common until the 1920s. These energy ressources had to be purchased by cash out of the surplus the farm could produce. Surplus was scarce, and as a consequence purchases were minimal. As a result the farm was an almost self-sufficient entity exchanging very little with the rest of society. In one single area however a service from the outside was unavoidable. That was the milling of the grain. This kind of work demanded more motive power, than muscles of horse and man could supply.

The growth of agricultural production from 1860 to 1914 was made possible by increasing energy consumption. At many farms, power for the threshing machine and the chaff cutter was provided by the horsewalk, whereas larger farms introduced steam power from the 1860s. The steam engine was too expensive to evict the horsewalk from minor farms. The years of depression during the 1870s and the 1880s were conducive to a change in Danish agriculture towards greater export orientation. Production gradually changed from grain to dairy products, bacon and eggs. This change proved very successful during the general European prosperity from the mid 1890s to the First World War, when export prices of butter and bacon doubled, and total animal production increased by about 70%.

This expansion of production demanded significant investments in agricultural machinery.[7]

The renewal of Danish agricultural production could only be so powerfully accelerated because the farmers understood the significance of modern technology and organization. General education had reached a relatively high level. The folk high schools mushrooming during the second half of the century made them awake and extrovert, and motivated them to join the cooperative movement and political organisations. Cooperatives were created to purchase and run the equipment for large units of production or for mechanisation. The farmers' newly acquired skills made them able to introduce and spread new technologies more rapidly than ever before.

The demand for new means of agricultural production called for new kinds of motive power and lighting. Renewable energy sources attracted new interest in the last decades of 19th century. Small wind motors were installed at a number of farms, especially near the coast, where sufficient wind was available. In the 1880s the internal combustion engine was developed to a practically useful stage. Especially suited for rural areas was the kerosene engine. In the beginning of the 1890s this type of engine had been developed to fairly competitive forms. Still it advanced only slowly in agriculture. Kerosene engines suffered from several problems, due to difficulties in getting a perfect combustion. This resulted in blockings, malodours, starting difficulties and insufficient fuel economy. Besides that the start-up took at least 5-10 minutes. Finally the fuel should be fetched, stored and filled up at regular intervals.[8]

As mentioned earlier the new electric motor had many promising qualities compared to other kinds of motive power including kerosene engines. The same applied to electric lighting. Like gas engines the electric motors only used power during operation. Moreover, they did not take up that much room and could be placed almost anywhere. They demanded very little maintenance and problems with noise, smoke and air were absent. They were cheap to buy, and even very small motors and varying workloads made only few problems.[9] The electromotor could be used to drive the threshing machine, the chaff cutter, the grinding mill, the beet slicer, the band saw, the pump etc. The electric motor must have seemed very attractive and above all profitable for farmers at that time. Not many farmers expressed their opinion about electricity in print. A few farmers did in the journal of DVES. In 1906 Davidsen, a Danish farmer, expressed the following opinion:

> "As regards the usefulness and profitability of a farm power station, I have realized, after the short time I have used it, that it is by this way the Danish agriculture will get its share of the great advances in the field of technology. The up till now usable prime movers have so far only been taken into use on

big farms and only to some extent. Since these machines are bound to a certain place and cannot be set up and used in every corner, like an electromotor, which, mounted on a waggon, can within a few minutes be moved to the place, where you want it to be used, and used immediately at full speed. In this way much labour is saved, and consequently a power station can easily yield an interest; still its greatest advantage is the greater control over operations enabling you to carry out a more intensive operation and thus obtain a larger income."[10]

Electric lighting seemed very attractive, too. In 1907 Pedersen, another Danish farmer, expressed his opinion about electric lighting in the journal:

"the light, you get in your stables and barns from a lamp it is very poor and bad, even if you use kerosene, which is still not common. The oil lamps are even worse. Not so with electric light. And it stands up to be placed anywhere, you do not need to worry about fire risks. In your dining room you can already have good light from ordinary lamps. But you will not do without your electric lamps, when you first have them. The women avoid the daily cleaning and filling. And you can so easily put on and out your lamps without getting kerosene odour in your room."[11]

Evidently farmers were motivated for early rural electrification. The quotations above and other passages in the journal suggest that farmers wished electricity primarily for intensifying production, with the aim of improving profitability.

5. Implementing the electrification projects
The influence of la Cour and DVES

Poul la Cour and DVES wished to promote a rural electrification of Denmark. But did la Cour and DVES succeed to get influence? In other words: Did they get the farmers to do what they wanted them to do? One way to throw some light on this question, is to quote keen contemporary observers.[12] Another way is to examine if the spread of small rural power stations was linked to the successive stages of la Cour's and DVES' effort. If such a coupling occurred it is highly probable that the activities of la Cour and DVES influenced the development of these successive stages.

In the period 1891-1902 la Cour followed a strategy of innovation of wind-electricity. None of the attempts of creating small industrial applications of electricity for rural areas turned out successfully. No spread of electric technology occurred in this period. Then in 1902, as earlier mentioned, la Cour converted the experimental mill to a power station for the village of Askov.[13] Subsequently, in the period 1903-1916, DVES followed a strategy of spreading the knowledge of small DC power stations in rural areas. This campaign progressed concurrently with the spread of rural power plants. The general European economic prosperity had created better export markets for agriculture. This contributed significantly to the emergence of the neccessary optimism among farmers. Now farmers dared make investments in this sort of technology.

The first rural DC power stations worked as small centres for the spreading of electricity to neighbouring villages. The powerful spreading from the beginning of DVES in 1903 until the death of la Cour in 1908 is striking. The number of established power stations rised from zero to a level of about 30-40 a year in 1908. From 1908 until the beginning of the First World War, the number of newly established small power stations was approximately 30-40 a year. At the beginning of 1920s the number of power stations reached a maximum at 374. Most of these were organized as cooperative societies.

In continuation of la Cours R&D effort the emphasis was at the outset on wind-electricity. From 1902 la Cour began to propagate the idea of wind-electricity followed by the newly established DVES in 1903. Projecting of wind-electric plants was of greatest significance for DVES in the period 1904-1907. But the strong development of the internal combustion engines limited the possibilities of wind power. Here the earlier mentioned Danish machine factory Burmeister & Wain also played a part. At the end of 1903 the firm had developed a small diesel engine which could pull a DC dynamo. This engine was quickly delivered to the Danish market. The development of internal combustion engines greatly advanced the spread of small DC power stations. DVES gradually changed strategy. This was reflected in the earlier mentioned modifications of the aim of the society. Instead DVES increasingly involved itself in the spread of power stations based on internal combustion engines. From 1907-1909 installation of gas engines was of great importance for DVES. The years 1909-1912 were strongly dominated by installation of diesel engines. This tendency also occurred in the pattern of the spreading of small DC power stations.[14] This suggest that DVES succeeded to maintain its influence on this spreading throughout the period 1904-1912.[15]

DVES planned a total of 131 plants until it withdrew from projecting in 1912.[16] Compared with the maximum number of small rural power stations, 374 attained at the beginning of the 1920s, this suggests a significant influence of DVES. In the period 1904-1919 a total of 230 rural electricians was trained, instructing nearly 20 electricians a year. The majority found jobs as rural electricians, which also indicates that DVES exerted significant influence.

It is difficult to assess the amount of influence exerted by la Cour's and DVES' popular campaign for electricity. Much evidence suggests that a very large number of lectures were delivered to farmers. Several books were published on the subject and a big correspondence were conducted. This effort suggests a significant influence. The above mentioned contemporary observers all emphasized the great importance of the popular campaign for electricity.[17]

What conditions permitted la Cour and DVES to influence the actions of the farmers during early rural electrification? On the one side they were helped by the external circumstances. The general European economic prosperity since the 1890s allowed farmers to make investments. On the other hand la Cour and DVES had the necessary understanding of the trends of development in Danish agriculture. But they miscalculated the need for wind-electricity. They did not have the necessary insight in the trends of development of internal combustion engines. Especially not concerning the diesel engine. This error was percieved by DVES. The new insight was reflected in the above mentioned changed strategy, now stimulating the spread of power stations in general. This succeeded because la Cour and DVES now had got the necessary understanding of potentialities and trends of development in internal combustion engines.

DVES and la Cour as users of influence

Poul la Cour and DVES were influential. But did they succeed to use their influence to promote rural electrification or were they ineffective? In other words: Was rural electrification promoted more than a normal capitalist spreading would have done in cooperative rural Denmark at that time? For example might electrical power companies under such circumstances have exerted similar influences due to their normal marketing strategies. However wind-electricity was so unique for la Cour and DVES that electrical power companies could hardly have spread this technology in the country-side. But such companies certainly could have spread other relevant technologies like power stations based on the internal combustion engine. And maybe the campaign of DVES and la Cour was not that important compared to normal company advertising.

One way to throw light on the before-mentioned question is to investigate the marketing effort of these companies in Denmark and abroad. It seems however that la Cour's and DVES' activities made the marketing of the electrical power companies more rational and effective. The farmers did not have the necessary qualifications to make a rational choice between the many offers made by different electrical power companies. A trustworthy impartial advisory and projecting expertise was needed. DVES worked as that to avoid each company wasting money by sending engineers and technicians to the same potential buyer. However DVES withdrew from the projecting market in 1912, when competent engineering companies had grown up.[18]

The only way to give a complete answer to the above-mentioned question is to repeat early rural electrification in Denmark experimentally without la Cour and DVES. But this is of course impossible. Then we could compare the time of take-off of rural electrification in Denmark with other nations or territories with similar conditions, but without la Cour and DVES. No nations is exactly similar in this respect, but we can compare with nations, having much in common with Denmark. The proper take-off of a proper general rural electrification independent of supplies of electricity from towns and industries happened just before 1914 in Norway, about 1910-1912 in the Netherlands, in 1917-1919 in Sweden and not until 1919 in Great Britain.[19] Compared to these four countries Denmark was the earliest starter, as take-off was in 1902. In Norway and Sweden the cooperative movement also played a part. This may suggest that la Cour and DVES succeeded to promote early rural electrification more than a normal capitalist spreading would have done in cooperative rural Denmark at that time. But this is a subject in need of further research.

But if la Cour and DVES promoted early rural electrification they also promoted the starting of a further rural electrification. The spread of the basic knowledge of electricity all over the country-side stimulated the later introduction of high voltage AC.

What conditions permitted la Cour and DVES to use their influence to bring about the suggested earlier take-off for rural electrification? Except for the original idea of propagating wind-electricity they now had the understanding neccessary for realizing an earlier take-off. They were aware of their own possibilities of exerting influence on the farmers and of the deficiencies of the ordinary capitalist spread of new technology. This insight was used to make capitalist spreading more efficient than it else would have been. Through the earlier mentioned popular campaign, training of rural electricians and impartial guidance for construction of power stations.

Consequences for the farmers

Poul la Cour and DVES wished to improve rural economy by the spread of electricity. And the evidence suggests that farmers wished to use electrical equipment, with the aim of improving profitability. But did they succeed?

Technologies like the electromotor and electric light were flexible. They were efficient means to intensifying production. The suggested early start of Danish rural electrification might have given the involved farms an increased competitiveness on the export markets compared to non-electrified farms and dairies.

Poul la Cour also wished to stop or reverse the migration from country to towns by giving country and towns equal opportunities as to use of electricity. According to statistics, the migration from country to towns, which had been quite considerable during a twenty year period, decreased somewhat in the decade following 1906. Then it increased again during the First World War, after which the situation stabilized in the 1920s. Unfortunately it is impossible to isolate the impact from rural electrification on this development. Other important impacts are too many. On the one hand electric machinery were labour saving devices and may have increased unemployment in the country-side. This may have stimulated the migration from country to towns. On the other hand the use of electrical equipment may have improved the economy of the farm by a more intensive production. This may have resulted in increased employment in rural areas. About the net effect we can only guess.

Furthermore it is difficult to determine if and how much early rural electrification improved the economy of Danish agriculture. The impact on economy drown among many other more important impacts like the good export marketing possibilities, other technological developments, improvements in organisation of farming, government policy etc. Again the only way to give a complete answer is to investigate what would have happened to economy if electricity was not introduced.

One way to say something about the effect on rural economy is to look at the indirect consequenses. As earlier mentioned the number of power stations reached a maximum at 374 at the beginning of 1920s. Afterwards the number went down a little, but sustained over 300 until the Second World War. Later during 1950s and the 1960s all small power stations were closed. This very slow decline was caused by the spread of a new means to supply electricity. The further general electrification of rural areas was carried through with high voltage AC. During the first decade of the 20th century high voltage AC technology was satisfactorily developed. Compared with small DC power stations high voltage AC power could cover large areas and costs of production could be kept low because of economies of scale. In the years 1910-1914 farmers began to get interested in AC power. Apart from the bigger villages, where electricity prices were almost the same as in smaller provincial towns, the electricity price of the small rural DC power stations was too high. Hence in the long run small DC power stations could only supply a part of the rural population in an economical way. Instead farmers preferred high voltage AC power stations supplying the surrounding districts. In several places a cooperative society supplying a certain area was etablished. The long preservation of the small power stations and the strong motivation to continue by installing of high voltage networks suggest that early rural electrification satisfied the economic expectations of the farmers.

What conditions permitted the farmers to improve their economy by the aid of electricity? On the one hand, as mentioned above, they were helped by the general European economic prosperity. On the other hand they had the necessary understanding for doing this. They realized the significance of electricity and how to organize it properly. They had the necessary qualifications for doing so. As already mentioned the folk high schools taught the pupils to appreciate innovation and cooperation. The farmers' newly acquired skills made them able to introduce and spread new technology rapidly in the rural areas. Especially they made them able to use the assistance from la Cour and DVES.

In 1907 Esbensen, another farmer, expressed the following point of view in the journal of DVES:

> "high voltage will not reach every corner of the country for a long time; until then small and large private installations, in the shape of farm power stations or cooperative undertakings will have done a good pioneer work for alternating current. ... Surely it can rightly be argued that pupils of the Danish folk high school and the agricultural schools have been the mainstay of the well-run Danish farm and smallholding ... It is likely, that the pupils of these schools may open a passage, first for the small power stations, later for the utilization of bigger ones in the practical life of the country; surely they are having a double task of throwing "light on the country"."[20]

Notes

1. J. Rasmussen: "Energien til magten", AUC 1987, p. 39-53; V. Faaborg-Andersen: "Denmark's agriculture and electrification", p. 494-496, in "Transactions, Third World Power Conference" vol. VIII, Washington 1938; V. Faaborg-Andersen: "Elektrificeringen" in "Danmarks kultur ved aar 1940" vol. IV, Copenhagen 1942 p. 38-58; S. B. Böcher: "Danmarks elektrificering" in "Geografisk Tidsskrift" no. 47, 1944-1945, p. 4-20; B. Wistoft, F. Petersen and H. M. Hansen: "Elektricitetens Aarhundrede", Dansk elforsynings historie, vol. I, 1891-1940, 1991; R. Henriksen: "Samarbejde mellem elektricitetsværker" and A. Ebbesen: "Elektriciteten i landbruget" in C. E. H. Dahl and V. Faaborg-Andersen (ed.): "Elektricitetens historie", Copenhagen 1939; N. J. D. Lucas: "The role of institutional relations in Danish energy policy" in "The journal of the David Davies Memorial Institute of International Studies" vol. VI no. 2 nov. 1978 p.347-353; A. K. Bak: "Elforsyningens udvikling 1911-1961" in "Danmarks Tekniske Museum, Årbog" no. 9, 1961 p. 62-69; A. R. Angelo: "Elektricitetsforsyningens udvikling i Danmark" in "Elektroteknikeren" no. 21, 15-11-1928, p. 471-478.
2. H. C. Hansen: "Poul la Cour – grundtvigianer, opfinder og folkeoplyser", Askov 1985; J. Th. Arnfred: "Poul la Cour som opfinder" in "Danmarks Tekniske Museum, Årbog" no. 16, 1968, p. 11-40

3. O. Hyldtoft: "Med vandkraft, dampmaskine og gasmotor. Den danske industris kraftmaskiner 1840-1897" in "Erhvervshistorisk Årbog" 1987, p. 116-118.
4. The following passage about la Cour is mainly based on H. C. Hansen, op. cit.
5. R. Skovmand: "Poul la Cour i nyt lys, H. C. Hansens la Cour-biografi", p. 11, in "Dansk Udsyn" no. 5, 1985.
6. P. la Cour (ed.): "Tidsskrift for Vindelektrisitet", january 1904 p. 3, march 1904 p. 31, november 1905 p. 184, november 1906 p. 282, january 1908 p. 401, june 1913 p. 587, march 1914 p. 635-638, march 1916 p. 711.
7. O. Hyldtoft: "Et rids af den teknologiske udvikling i Danmark", p.12-13 in "Teknologisamfundet 1984 – en debatbog", 1984; S. P. Jensen: "Nye maskiner, redskaber og energiformer" in C. Bjørn (ed.): "Det danske landbrugs historie III, 1810-1914", 1988, p. 266-275.
8. O. Hyldtoft 1987, op. cit., p. 105-116; S. P. Jensen: "Nye maskiner, redskaber og energiformer" in C. Bjørn (ed.): "Det danske landbrugs historie III, 1810-1914", 1988, p. 266-275.
9. O. Hyldtoft 1987, op. cit., p. 105-116.
10. P. la Cour (ed.), op. cit., may 1906 p. 223.
11. P. la Cour (ed.), op. cit., january 1907 p. 305.
12. H. C. Hansen, op. cit., p. 369-372, 375, 416.
13. The experimental mill was the first wind power station in Denmark, but strictly speaking not the first village power station. Among the stations existing in 1922, three were established before the experimental mill.
14. H. C. Hansen, op. cit., p. 367.
15. P. la Cour (ed.), op. cit., june 1913 p. 589.
16. P. la Cour (ed.), op. cit., june 1913 p. 587-591.
17. See note 12.
18. P. la Cour (ed.), op. cit., january 1904 p. 1-2, january 1908 p. 397-399, june 1913 p. 587, march 1914 p. 635-637, march 1916 p. 711.
19. L. Thue: "Hvorfor ble Norge et rikt land? Lokale kooperasjoner og økonomisk vekst"; G. Nerheim: "Patterns of Technological Development in Norway" in J. Hult and B. Nyström (ed.): "Technology & Industry. A Nordic Heritage", USA 1992; S.-O. Olsson: "Tröskan, ångan och elen. Jordbruksteknik i seklets börjarn" in "Polhem", 1987 no. 2; The chapter "Rural Electrification" in "Transactions, Third World Power Conference" vol. VIII, Washington 1938.
20. P. la Cour (ed.), op. cit., january 1907 p. 309.

The Development of German Engineering Education in the Nineteenth Century – a Comparison with Great Britain and France

Klaus Mauersberger

To follow the development of institutions of engineering education in the course of the 19th century would just as well mean following nearly endless and indeterminable lines and angles of development, interactions and backgrounds. This paper will, while referring to Peter Lundgreen and others[1], who have treated aspects of education and professionalization, examine the theoretical-methodological principles of education in particular as well as the scientific conceptions of teaching plans, their origins and their ways of maturing. The paper will focus mainly on mechanical engineering, which was to a much larger degree a field of private enterprise. Civil engineering, in contrast, was primarily dominated by state engineers.

It may seem reasonable to treat these aspects within an all-European framework, but in this context it will be sufficient to concentrate on the main trends of those countries, which determined the overall development.[2] Characteristic statements can be made for the German-speaking region of the mid-19th century without generalizing beyond meaning. Of course, quite a few areas of engineering history are still in a state of flux, and research into the history of engineering science has not yet yielded results, which allow us to draw wide-ranging conclusions which concern both cognitive and social aspects equally.

There is no doubt that the renowned Parisian *École Polytechnique* formed the cradle of modern engineering education. The innovative impulses of this nursery of the engineering sciences can, however, only be understood from the scientific tradition of the ancien régime. The important military colleges and schools for the élite preliminarily shaped the later engineering education and thereby made up a continuing line between the engineering corps and the later polytechnical schools.[3] Nearly all the founding fathers of French polytechnical education were rooted in these forerunner institutions, e.g. the great *Bernard Forest de Bélidor* (1693-1761), who worked at the artillery school at Le Fère.

Gaspard Monge (1746-1818) and *Lazare Carnot* (1753-1823), who heavily influenced the training programmes, came from the *École du Génie Militaire* in Mézières, which was established in 1748. Also *Gaspard Riche de Prony* (1755-1839) was educated at the famous *École des Ponts et Chausées*, the Parisian school of roads and bridges, which had been formed in 1747. These schools largely made up the first monotechnical colleges in the world.

An analysis of the educational concepts as propogated by the École Polytechnique has to be based on an understanding of the French background. It must be appreciated that this prototype of future polytechnical schools originated from the military requirements of the French Revolution and the central state of France. It was directed at academic training of state officials for public service, with military and civil engineering as major aims.

The industrialization of France does seem extremely slow on this background. At the beginning af the 19th century, France had still not fostered any industrial machine building industry. Later on it was exactly this industry which gave a great impetus to the educational profile of the engineering sciences in Germany.

Moreover, it must be emphasized that the École Polytechnique marked the culmination of a long discussion on the theory of education, which went on during the 18th century. This discussion was part of the Enlightenment and was continued by encyclopedists and physiocrats. An orientation towards application and democratization emerged as its leading idea at the close of the century. The pedagogical idea of the formation of reality was to be complied with mainly by laying emphasis on the natural sciences as a means for educating the mind. According to this educational concept, the establishing of a real equality between citizens required an active attitude towards knowledge by the learner. This included the capability to develop solutions to problems by theoretical means.

What was the educational programme like in detail? In general, the French polytechnics were guided by the unity of the engineering education. The unity of all branches of engineering knowledge was expressed in the curriculum of the joint preliminary education of the École Polytechnique, which was directed considerably towards the imparting of engineering fundamentals. Physics (mechanics) and chemistry were regarded as the main disciplines. Above them stood mathematics as the connecting element and principle of applicability. We will return to this subject later on.

Specialized education, however, was imparted at the socalled *Écoles d'Application* in a strict bipartition. These schools were in the direct line of tradition as laid down by the special schools and the engineering corps already mentioned. An alteration between theoretical studies and pratical construction exercises was used. As a rule, the students were assigned to the building authorities or mines in the *departements* to carry out practical work during the summer months.

Essential educational principles of the École Polytechnique were incorporated into the curriculum of the "geométrie descriptive" by Gaspard Monge. Descriptive geometry constituted the basic curriculum of the institution after 1795, and we must therefore deal with its concept in a few sentences. At first glance it may seem odd that the curriculum of one single discipline was raised to form the organizing didactic principle of the whole institution. This, though, can be understood only from a situation characterized by a striving for an advanced engineering education while at the same time the theoretical and methodological conditions required for the very same education still had to be created.

All in all, a new form of relationship found its expression in Monge's interpretation. His dynamic and application-oriented understanding of science was directed at the discernibility and control of natural processes for the benefit of human society. Monge regarded quantification as the essential prerequisite for a practical exploration of technological processes. His general methodology, however, was directed not only towards the concept of quantity but also towards the specific idea of a "law of nature cognition", which was adopted from the natural sciences. Epistemologically, Monge started from an orientation of knowledge to objects according to areas of application. Theoretical findings included forms and motions of bodies, their composition and internal structure as well as properties. Besides the "intellectual sciences" such as geometry, mathematics and physics, Monge regarded empirical fundamentals as important. These were acquired through drawing exercises and chemical experiments, which was quite a novelty at the time. In Monge's opinion descriptive geometry ought to have an integrating function as a means of communication and as a research method. As the language of science it was to link the artisan – "not having a language" – with the natural scientist.[4]

What, then, were the special features of the method proper? Whereas the Cartesian geometry had fully served the illustration of the mind, now the illustration and documentation of technical objects was on the agenda. The general rules and theories of descriptive geometry allow the mapping of spatial configurations in two-dimensional drawings. Compared with the analytical representation which prevailed in mechanics, which was not excluded, the method was characterized by a sensorial-descriptive and therefore consequently didactic representation. It opened up a wide field for application to problems connected with machines. An example of this is the kinematic theory of the Monge-school by *Hachette*, *Lanz* and *Betancourt*, in which an anticipatory concept of design in the mechanical sciences took shape.

The possibility of deriving connections related to construction and manufacturing from drawing was highly important for the synthesis of new technical means. Moreover, aims of training were the teaching of accuracy and draft-related thinking. Exactly this precision and reproducability were demanded by the arising mechanized production. The well-directed influence necessitated a language by means of which the connections between shape and function were made clear. By this concept engineering drawing, though its roots go way back historically, and the geometrical representation of technical processes of motion, was raised to a language of the engineer.[5]

In summary, the descriptive geometry, having fully developed from didactic questions, became the determining foundation of the system of engineering education and its corresponding theory of science. Its synthesizing character is obvious from the affinity with graphic technique. If we ask whether a basic methodological framework such as this could meet the demands of practice, we would have to answer in the negative. Initially, the educational value prevailed. A change in the teaching of technical knowledge was gaining ground, finally to influence the whole of Europe. In France the methodological prerequisites for the modern engineering sciences were created. The further development of educational institutions in the Napoloenic era can be reduced to the denominator "bureaucratization and militarization of education". Whereas initially descriptive geometry still constituted a link to the "uneducated" technician, at the École Polytechnique opinions soon spread, which once again overwhelmed geometry by omnipotent analysis. Under the auspices of the renowned celestial mechanic-scientist *Pierre Simon Laplace* (1749-1827), the educational concept tended to lay emphasis on purely theoretical subjects. Teaching gradually lost its orientation towards practice. This could be compensated for only partly by the educational profile of the special schools.

This trend was not at all due to the unfulfilled expectations of the "géométrie descriptive", but it reflected caracteristic features of the educational political situation on the threshold of the industrial revolution. The general stagnancy became a hindrance to the engineering education. Inevitably, considerable drops in the motivation to study took place. The registration and examination system expanded into an independent apparatus. In other words, the École Polytechnique "degenerated into a swotting institution". The "École de Monge" had given way to an "École de Laplace".

It was inevitable that this scientific singlemindedness nurtured considerable criticism on the part of those who continued advocating a unitarian principle. Moreover, during the thirties conditions by which industrialization gained more and more momentum had matured in some branches of production. It is no wonder, therefore, that amongst the critics of the prevailing system of engineering, especially the mechanical scientists insisted on a reorganisation of this education. Renowned scientists such as *Arago, Olivier, Dupin* and *Poncelet* observed new educational trends emerging abroad, especially in Germany. As a reaction to the needs of the private capitalist sector, the *École des Arts et Manufactures* was established in Paris in 1829 as a private foundation. The mechanical sciences in France were thereby freed from the role of being an appendix to the schools of mining and civil engineering. Neither the role played by the *lycées* and the private *Écoles Prepatoire* could be overlooked, or the many-sided effects of the Conservatoire des Arts et Métiers with its technical collections for the training of skilled workers and foremen.

The new methodological concept which made way between 1830 and 1850 was supported by important engineering scientists such as Jean Victor Poncelet, Artur Morin, Regnault, Charles Dupin and others. Basically, it consisted of a combination of the analytical and the methodological concept.

Poncelet's "industrial mechanics" evolved as the dominant discipline of this paradigm. Graphic analogies and approximate methods became methodologically important only through the combination of precise theoretical prerequisites with practice-related empirical methods. Out of this grew the exemplary effects of the French concept in the following period. Finally, the transition from geometry to mechanics had been completed.

Before turning to Germany, we must take a look at England, which – as far as the developments in engineering and of the productive forces are concerned – held the first rank. Let us then shortly follow the development in Great Britain, at least from the beginning of the Industrial Revolution. The radical technological and social changes on the British Isles took place within the framework of a natural utilitarianism which grew out of British Puritanism and Liberalism. Mechanical engineering, for example, which had still not separated itself from its artisan background before 1800, rose on the basis of a long established empirical experience. For training, the workshop principle – the socalled "shop culture" – prevailed. Although the engineering sciences stabilized themselves in the course of increasing problems in engineering practice, the deep-rooted English liberalism prevented state interference with respect to their institutionalization. Thus, in England the task of educating engineers was shared by civil engineers, many of them working scientifically and and carrying out experiments, while also often being entrepreneurs themselves, private foundations, societies, patronages and institutions as well as the applied research carried out at the universities. This was highly important, but carried no social prestige.

During the first decades of the 19th century the situation on the continent was characterised by the fact that early industrial England surpassed its European competitors completely. On the British Isles there was no hint that engineering sciences were to take on a much more prominent position. This was due to the relative prosperity based on entrepeneurial abilities and to the pursuit of profit, which could be satisfied rapidly as well as to an absorbing market. In the financially relatively weak areas, especially the countries of the German-speaking region, the factor of engineering education attained an important position in supporting and promoting trade and industry. Moreover, together with technological progress which was somewhat delayed, a need of the skills of the engineering profession developed, which was very much different from the English development during the early phases of industrialization. Heavy competiton and tutelage as well as a considerable retardation forced an accelerated technological and industrial development. An appropriate means for achieving this should have been the establishing of a well functioning engineering education. With greater theoretical clarity and mental flexibility, such an education would possibly be able to match the English in areas in which they were superior to their European neighbours.[6]

Despite the exemplary effect that emanated from France in the field of engineering education, the Germans did not simply copy the French pattern. The possibility to do so was ruled out by the fact that there was no central state. In a configuration of territorial states, engineering education had to unfold in much more varied forms. It is therefore difficult to find the lowest denominator. Moreover, with the exception of civil engineering, right from the beginning there were no restrictions imposed on technical civil service as in France.

The backgrounds of the different technical schools cannot be dealt with any further in this paper. Rather, attention will be drawn to subjects such as content and methodology. As expressed by the term "polytechnical", excluding the monotechnical institutions such as mining academies, which still existed, the supreme requirement was the unity of engineering education. This trend exeeded that of France, since the unity of basic and specialised training in one single school was what was strived for. The Prussian model was an exception. In Berlin, the civil engineering school and the mining academy as well as the higher trade institute (*Gewerbeinstitut*) co-existed, the latter most likely based on polytecnical conceptions.[7]

The first teachers at the technical schools were polytechnicians in the truest sense of the word. The professional university of *Johann Andreas Schubert* (1808-1870) from Dresden or *Julius Weisbach* (1806-1871) from Freiberg were proverbial. At a time when mechanical engineering in Germany was still getting ready to move from copying English prototypes to independantly developed constructions, exactly this versatility was in demand. This is reflected in various manuals and other similar practical books. It is certainly not an exaggeration to speak of a paradigm of recipes and relation-factors (Verhältniszahlen) behind the development of scientific-mechanical engineering.[8]

Mechanical technology or manufacturing was still in its early descriptive stages. Chemical technology had been raised to a university subject, but there it was being gradually replaced by pure chemistry. At the polytechnical institutions it was generally taught according to scientific principles. Getting a chair at a polytechnical school to many a chemist meant a springboard for a chair at the university. Distinguished chemists such as *R. Bunsen* (1811-99) and *F. Haber* (1868-1934) used this path to the university and at the same time heightened the standard of the natural sciences in engineering education. In the founding phase of the polytechnics, the highest degree of scientific penetration was found in civil engineering, where use was made of the high standard of French construction mechanics.[9]

It is difficult to point out a common methodological denominator in this variety of polytechnic institutions. A denominator was seen only in programmatic statements, of which the "peculiar technological method" ("eigentümliche technische Methode") of *Prechtl* in particular deserves mentioning. Prechtl was the spiritual father and director of the *Wiener Polytechnisches Institut* for many years.[10] His method was based on the concept that the creation of engineering was was neither "applied natural science" nor "pure empiricism", but needed a specific approach. The methodological determination of the relation between natural-scientific fundamentals and practical experience or the tentative struggling for this aim continued to be a main component in the cognitive process of the engineering sciences and in engineering education. It remained a problem, showing its face again and again on higher levels.

On the way to a "universitas scientiarum technicarum", there were quite a few schools of methodological thought, which on the concrete level were influenced decisively by personalities, regional peculiarities and scientific factors as well as socio-economic and education-political interests.

In the forties and fifties a considerable process of differentiation began in the developing of the engineering sciences, dictated both by momentum and diverse practical requirements. Although the polytechnical character remained a school principle, a principle of specialisation established itself within the institutions.

Civil engineering, mechanical engineering and chemical technology were still the three pillars of polytechnical education. In Karlsruhe, following a reform planned by *Ferdinand Redtenbacher* (1809-1863), a significant and rapid increase in the quality of teaching could be seen as early as the late 1850's. It was the development of mechanical engineering, which established the new profile. Redtenbacher, a great teacher-personality, attracted students in large numbers to Karlsruhe. His mechanical sciences after the French pattern determined the development of methods far beyond the borders of Baden. The 1850's constituted a great period for scientific mechanical engineering and machine building. The railway, the water turbine and the steam engine were all typical attributes of the industrial era.

Redtenbacher's creation – together with the proximity of France – also inspired the *Züricher Eidgenössisches Polytechnikum* (Swiss Polytechnic in Zurich), which is currently regarded as the first technical institution, for which a university level and character was scheduled. It was established in 1856. In the meantime, the demand for scientifically educated engineers came from the fast growing large-scale industry. This corresponded with the intentions of engineering scientists to consolidate and sophisticate the theoretical foundations and to raise the status of the polytechnicians by academic training. Their endeavours to furnish proof of a high scientific standard were connected with emancipatory efforts by the engineers, which occasionally led to some delimitation from the universities and the principles of neo-humanistic education. The engineer was still regarded as a parvenu and an intruder on established social hierarchies. Therefore, social and scientific tasks constituted a unity.

The plans of reorganization in the sixties followed a remarkable process of formation of opinion, which did not at all show agreement, because education remained too heterogenous. Especially the *Verein Deutscher Ingenieure* (the professional organisation of German engineers) with its most competent advocate *Franz Grashof* (1826-1893) propagated the university concept. In his notable lecture "On the principles underlying the organisation of polytechnical schools" in 1864,[11] the energetic pioneer of a university-level engineering education demanded the separation of a medium-level education from the higher engineering education as well as the establishing of the so-called "General Departments for Alignment with the Educational Standard of Universities". Already in the following year Karlsruhe got a new organisational statute. The status of technical college was conferred in Munich in 1868, in Dresden in 1871, in Brunswick in 1877 and in Berlin in 1879 with the merging of the building academy with the higher trade school.

It must be emphasized that content-related and institutional reforms in engineering education were continually carried out, but nevertheless, peaks of general rapid increase in quality were experienced. Most situations like these were characterised by increased travelling by the responsible scientists and by the publication of a considerable number of memorandums. During the discussion on reorganisation in Austria after 1848, the polytechnic professor *Karel Koristka* (1825-1906) of Prague travelled through the European countries in 1853 and 1862, looking for model concepts for the reform of the technical schools in his country.[12] Forty years later, *Alois Riedler* (1850-1936), the militant advocate of a further reform movement, already looked at the American institutions,[13] and the Prussian Ministry of Education and the Arts sent Dr. *W. Lexis* on a travel through Europe and the USA in order to form an opinion about the existing level and the prospects of engineering education.

Let us now return to the polytechnics, which had been promoted to colleges. Thanks to outstanding scientist personalities, who gave a particular profile to their institutions, at the same time a broad methodological variety was created. The scientific character of these institutions was decisively shaped by scientists such as the civil engineer and founder of graphic statics, *Carl Culman*, as well as the chemical technologist *Georg Lunge* (1839-1923) in Zurich. Others were the materials scientist Bauschinger and the refrigeration engineer *Linde* in Munich, the theorist of the strength of materials *Bach* in Stuttgart, the mechanical engineer *Karmasch* in Hannover, *Zeuner* in Dresden, who was working on technical thermodynamics, as well as the kinematics scientist *Reuleaux* in Berlin. All of these men endeavoured to provide guiding examples for instruction and research and to establish, in their own ways, a balanced relationship between the fields of activity assuming profile at the technical colleges or in the practice of engineers. Cases of one-sidedness and exaggerration did occur. The formation of scientific schools became pronounced especially in the German speaking region.[14] Exactly the often deplored territorial fragmentation led to competition among the schools in the individual countries or at least to the formation of a climate of advantageous tension between the independent communities of research and education. A differentiated image of teaching was also created by wide graduation of engineering education at non-classical secondary schools, technical institutes and trade schools as well as technical colleges.

In the meantime, some balance had also been established on an all-European level in spite of national peculiarities. The engineering educational system of England had followed suit.[15] The French schools had rather turned towards practice once again. The division of the educational system into three parts run through the European countries from the Netherlands and Belgium to Russia, albeit without taking on quite the same form. Non-classical secondary schools and lycées served the preparation for advanced establishments, to which in many cases they were attached. The lower engineering education was characterized also by drawing courses or Sundays schools, which admitted workers free of charge.

At the intermediate level, so-called trade drawing schools developed (in England *Schools of Design*, in France *Écoles du Dessin*), which raised the drawing courses to the level of being an independent subject. New methods in drawing instruction were introduced by *Haindl* in Munich and *Dupuis* in Paris. The so-called "Schools of Industry" in Belgium and Switzerland, the above mentioned Conservatoire in Paris and the *Mechanics Institutes* in England all belonged to the medium-level schools for artisans and tradesmen. Although the organizational structure and the financial conditions of these institutions differed widely, they had one feature in common: instruction in drawing was in the foreground.

Collections of models, mechanisms, machines and drawings were available to help put some life into teaching.

Also at some advanced technical schools, collections and workshop training were to be found. These collections had, however, more the character of specimen collections, and pronounced laboratory training was carried out only in chemistry. It is quite instructive that in France the private institutions, which tried to meet the demand of the private industrial sector, associated with the "grand écoles". Private institutions, it must be noted, were supervised by the state (how could it be otherwise in France ?). At the universities in England and Scotland real polytechnical experts such as *William Macquorn Rankine* (1820-1872) were to be found longer than elsewhere. After 1850 the first schools of the engineering sector were gradually established. On the suggestion of Professor *Lyon Playfair*, the *Government Schools of Mines and of Science applied to the Arts* was called into existence in 1851. The most eminent representative of the technical mechanics in England, *Robert Willis* (1800-1875), worked here. Advanced engineering education was furthermore imparted by the private *Queen's Colleges*.

Finally, let us now take a look at the last quarter of the 19th century. Trends in the engineering sciences during this period, concerning instruction and research, has been called a "struggle about methods".[16] We will try to characterize the new conditions for the training of engineers and engineering scientists.

The tendencies towards a unification of advanced engineering education, which had been advocated mainly by the *VDI* (Society of German Engineers), were given considerable weight by the economic pressure of the following years. In spite of the undeniable and important success of the technical educational system, the establishing of adequate relations between theoretical and empirical knowledge was proved only partly successful. The gap between teaching and the rapidly progressing industrial practice increased in many fields. Intensification and concentration of production as well as capitalist mass production pressed for a reorganization of academic life. A strong trend towards consolidation of the theoretical basis stood in the way for a development of disciplines in traditional areas. The scientific disciplines were grouped around "young industries" like electrical engineering, precision mechanics and optical engineering as well as the chemical industry, thereby gaining access to practice more rapidly.

Around 1875 methodological controversies started mainly with Franz Reuleaux's concept of engineering kinematics, which theoretically as well as methodologically was ahead of its time. Its practical value was questioned. Where today kinematics is regarded as the boldest heuristic conception of the engineering sciences at that time, the machine-building industry at the close of the century demanded a more application oriented approach. Disputes about methodology, which sometimes were held vehemently, soon spread from the theory of machines to other disciplines which had also descended into theoretical single- or narrow-mindedness. This was even part of the well-known discussion on "shop culture versus school culture". Efficiency in design, production and operation became an indicator of an adequate apparatus of theories and, thus, a regulating factor for proximity to practice once again. To express this in an up-to-date way, in the last decades of the 19th century a change of paradigms became apparant in some branches of the engineering sciences. This change was initiated by a broad discussion about educational principles, about overstressing theoretical methods.

In his article "Die Ziele der technischen Hochschulen" (The aims of the technical colleges)[17] in 1896 Alois Riedler, the militant protagonist of the engineers' movement, submitted a plan which was mainly directed at laboratory instruction. It was modelled on the American system and the flexible complying with the requirements of industry.[18] The development of an efficient materials testing and mechanical laboratories started c. 1870. Around 1900 the polytechnical schools which had been upgraded to the status of technical colleges were given a modern form by those institutions.[19] The complex experiments of the engineering sciences made up an essential contribution to the re-establishing of the unity of instruction and research, of basic education and engineering practice. In this way a balanced methodological concept emerged; a concept, which was able to meet the demands of all sectors. The long process of the engineers' self-understanding and emancipation reached a climax when the technical colleges were granted the rights to confer the doctorate.[20]

Finally, the following conclusions can be drawn from the turbulent developments at the close of the nineteenth century:

1) Methodological struggling is no independent process, delimited within the sciences, but an expression of many external and internal factors of development.
2) The conflicts did not involve all spheres of the engineering sciences.
3) The controversy was most intense in the forefront of research; moreover, in teaching and research, balancing concepts existed at all times.
4) The heated methodological debates were overlapped and intensified by the engineers' movement for professional emancipation.
5) A correlation existed between raising the status of institutions and the designing of ambitious curricula.
6) Theoretically independent trends, also single-tracked ones, can be comprehended through the dialectics of the development of cognition. Sometimes they constitute a necessary component of scientific consolidation and sophistication.

7) Experimental research and laboratory training on a large scale resulted from the involvement of the sciences in the conditions for utilizing capital.
8) For completion of the development dealt with in this paper, the regulating hand of the state was required.

Notes

1. P. Lundgreen, Engineering Education in Europe and the U.S.A., 1750-1930: The Rise to Dominance of School Culture and the Engineering Professions, *Annals of Science*, 47 (1990), pp. 33-75.
   U. Troitzsch, Technisches Schulwesen, Wissenschaftsorganisation und Wissenschaftspolitik in Deutschland 1850-1914, *Technikgeschichte*, 42 (1975) 1, pp. 35-43 (literature-report).
   F. Schnabel, Die Anfänge des technischen Hochschulwesens, *Festschrift anlässlich des 100-jährigen Bestehens der Technische Hochschule Friedericiania zu Karlsruhe*, Karlsruhe, 1925, pp. 1-44.
   W. Treue, Die Geschichte des technischen Unterrichts, *Festschrift zur 125-Jahr-Feier der Technischen Hochschule Hannover 1831-1956*, Hannover, 1956, pp. 9-60.
   W. Pfuhl, *Die allgemeinen Ursachen für die Entstehung des technischen Bildungswesens in Deutschland und die Einordnung der polytechnischen Schule zu Dresden in das System der technischen Bildung zur Zeit der industriellen Revolution*, (Ph. D. dissertation), TU Dresden, 1972.
   K. H. Manegold, *Universität, Technische Hochschule und Industrie. Ein Beitrag zur Emanzipation der Technik im 19. Jahrhundert unter besonderer Berücksichtigung der Bestrebungen Felix Kleins*, Berlin, 1970.
   H. Blankerts, Bildung im Zeitalter der grossen Industrie. Pädagogik und Ausbildungsgegenstand, *TU Berlin – Dokumentation*, 1(1979), pp. 1-30.
   M. Kranzberg (ed.), Technological Education – Technological Style, San Fransisco, 1986.
   J.H. Weiss, The Making of technological Man, Cambridge MA, London, 1982.
2. Peter Lundgreen, Bildung und Wirtschaftswachstum im Industrialisierungsprozess des 19. Jahrhunderts: methodische Ansätze, empirische Studien und internationale Vergleiche, Berlin, 1973.
   R. Locke, Industrialisierung und Erziehungssystem in Frankreich und Deutschland vor dem ersten Weltkrieg, Historische Zeitschrift, München, 225, 1977, 2, pp. 265-296.
   E.A. Musson (ed.), Wissenschaft, Technik und Wirtschaftswachstum im 18. Jahrhundert, Frankfurt am Main, 1977.
   G. Ahlström, Engineers and Industrial Growth, Higher Technical Education and the Engineering Profession during the 19th and early 20th Centuries: France, Germany, Sweden and England, London, 1982, pp. 173-76.
   R.A. Buchanan, Technological Revolution in East and West, Polhem, Stockholm, 3, 1985, 2, pp. 79-93.
   K. Mauersberger, Vergleichend Betrachtung zur Herausbildung der Mechanik und Maschinenlehre in England, Frankreich und Deutschland bis zur Mitte des 19. Jahrhunderts, NTM-Schriftenr. Gesch. Naturwiss., Technik u. Medizin, Leipzig, 22, 1985, 2, pp. 25-32.
   G. Zweckbronner, Ingenieurausbildung im Königreich Württemberg, Stuttgart, 1987.

3. F.B. Artz, The Development of the Technical Education in France 1500-1850, Cambridge MA, London, 1956, C.S. Gilmore, Coulomb and the Evolution of Physics and Engineering in Eighteenth-Century France, Princeton, N.Y., 1971.
4. M. Paul, Gaspard Monges "Geométrie Descriptive" und die École Polytechnique – Eine Fallstudie über den Zusammenhang von Wissenschaft- und Bildungsprozess, Ph. D. Dissertation, Uni Bielefeld, 1980.
5. A. Neduluha, Kulturgeschichte des Technischen Zeichnens, Blätter für Technikgeschichte, 19, 1957.
6. L.-U. Scholl, Ingenieure in der Frühindustrialisierung, Göttingen, 1978.
7. K.-H. Manegold, Eine École Polytechnique in Berlin, Technikgeschichte, 33, 1966, 2, p. 182.
8. K. Mauersberger, Die Herausbildung der technischen Mechanik und ihr Anteil bei der Verwissenschaftlichung des Maschinenwesens, Dresdner Beiträge zur Geschichte der Technikwissenschaften, 2, 1980, pp. 1-52.
9. Th. Hänseroth, Die "Mechanik der Baukunst" – zum 150. Todestag des Begründers der Baumechanik L.M.H. Navier (1785-1836), NTM, Leipzig, 22, 1985, 2, pp. 33-41.
10. A. Bihl, Aus der Geschichte des deutschen technischen Hochschulwesens im alten Österreich, Technikgeschichte, 30, 1941, p. 164.
11. F. Grashof, Prinzipien der Organisations der Polytechnischen Schulen, Berlin, 1866.
12. C. Koristka, Der höhere polytechnische Unterricht in Deutschland, in der Schweiz, in Frankreich, Belgien und in England, Gotha, 1863.
13. A. Riedler, Amerikanische technische Lehranstalten, VDI-Zeitschrift, 38, 1894, pp. 405, 507, 608, 629.
14. See e.g. K. Krug, Zur Herausbildung der Technischen Thermodynamik am Beispiel der wissenschaftlichen Schule von G.A. Zeuner, NTM, Leipzig, 18, 1981, 2, pp. 79-97.
15. G. Roderick & M. Stephens, Science and Technology at English Universities and Colleges on the Economic Development during the 19th Century, Technikgeschichte, 40, 1973, 3, pp. 226-250.

R.A. Buchanan, Institutional Proliferation in the British Engineering Profession, Econ. Hist. Rev., 38, 1985, 2, pp. 42-60.

R.A. Buchanan, The Rise of Scientific Engineering in Britain, British Journal for the History of Science, 18, 1985, pp. 218-233.

16. H.-J. Braun, Methodenprobleme der Ingenieurwissenschaft 1850-1900, Technikgeschichte, 44, 1977, 1, pp. 1-18.
17. A. Riedler, Die Ziele der Technischen Hochschulen, VDI-Zeitschrift, 40, 1896, pp. 301, 337, 374.
18. J.K. Finch, Engineering and Science: A Historical Review and Appraisal, Technology and Culture, 2, 1961, pp. 223-325.

    M. Kranzberg & C. Pursell, Technology in Western Civilization, N.Y. 1967.

    E.T. Layton, Mirror-Image Twins: The Communities for Science and Technology in 19th Century America, Technology and Culture, 12, 1971, pp. 62-80.

    E.T. Layton, American Ideologies of Science and Engineering, Technology and Culture, 17, 1976, pp. 696-701.

    M. Fores, Transformations and the Myth of "Engineering Science": Magic in a White Coat, Technology and Culture, 29, 1988,1, pp. 62-81.

    E.T. Layton, Science as a Form of Action: The Role of Engineering Sciences, ibid. pp. 82-97.

    D.F. Chanell, Engineering Science as Theory and Practice, ibid. pp. 98-103.

19. K. Mauersberger, Zur Entwicklung des Experimentalswesens in der Technikwissenschaften mit

besonderer Berücksichtigung der Materialprüfung innerhalb des Wissenschaftlichen Maschinenwesens, Dresdner Beiträge zur Geschichte der Technikwissenschaften, 10, 1984, pp. 68-103.
20. K.-H. Manegold, Technische Forschung und Promotionsrecht, Technikgeschichte, 36, 1969, 4. K.-P. Meinicke, Dokumente zum Promotionsrecht der Technischen Hochschulen in Deutschland (1899), NTM, Leipzig, 25, 1988, 1, pp. 79-82.

The Tension between Theory and Practice – Dutch Engineering Education in the Nineteenth Century

G. Verbong

1. Introduction

The development of engineering education in the last two centuries can generally be described – according to Lundgreen – as the rise to dominance of an academic education of engineers.[1] Lundgreen distinguishes two different patterns of development until 1870: a French-German one and an Anglo-American one. In France and Germany the academic training of engineers for functions in state bureaucracies started early. The role model of technical civil servants proved to be very influential on the engineering education and professions in these countries. In England and the USA state engineers were almost absent and the most important role model for the engineers was that of entrepreneurs or free advisors. In the field of private enterprise the training of engineers on the job stayed predominant. Because of the importance of civil engineering and the role of the state bureaucracy in this field, the Dutch case fits in general the French- German model. In this paper, I will describe the development of engineering education in the Netherlands in the 19th century. I will focus on three related themes.[2]

The first one concerns the start of engineering education in Holland. What were the expectations of the benefits of a more formal training of military and civil engineers? The first proposals for a polytechnical school were already made during the French regime (1806-1813), but it took nearly forty years before the Royal Academy at Delft started in 1842 with the training of civil engineers. In this paper, I will analyze which groups supported or opposed academic engineering schools and which arguments were used.

The second theme relates to the content of the education. The general opinion among the proponents of engineering schools was that the application of "the sciences on the practical arts" would be useful or even necessary for the development of the country. These ideas, however, had to be translated in a curriculum, which proved to be far more difficult than expected. Another problem was the lack of appropriate teachers to carry out the programme. In this regard, Dutch engineers, who studied at the Ecole Polytechnique and Ecole des Ponts et Chausses during the Napoleonic period and who consequently became educators at the engineering schools, were very influential in the shaping of engineering education.

The third theme is the relation between theoretical knowledge and practical demands. Ideas on the value of general scientific knowledge in practice influenced the foundation of engineering schools and the content of the curriculum to a large degree, but the graduates of the schools had to prove the value of their education in practice. Lundgreen explicitly states, that claims of academic engineers on the necessity of theoretical training were to a very large extent ideological or serving the self-interest of the engineers. What was the role of the employers of the engineers and did the experience in practice induce changes in the engineering education? In the analysis of Dutch engineering education I will focus on civil engineering, because civil engineers were dominant throughout the century.

2. Civil engineering before 1800

In 1600, during the independence war with Spain, Prince Moritz asked the University of Leyden to train engineers in fortification and surveying. The application of new Italian ideas on the structure of fortifications had proved to be very effective. Simon Stevin, a famous Dutch engineer and scientist, described and illustrated these ideas in a book in the Dutch language on fortifications.[3] He was asked by the Prince to design the curriculum. It consisted of elementary mathematics (arithmetics and geometry) [4], the study of various models of fortifications and surveying. The students acquired the practical experience in fortification methods usually in the army. For the training in surveying there was a kind of practice-area at the university, where the students could learn to use the instruments for measuring the land under various conditions. This course at the university of Leyden became generally known as the 'Duytsche Mathematique'. Several well-known professors in mathematics were appointed for this course. These lessons were, as the name already indicates, taught in the popular language, which was, of course, very unusual for a university in those days. However, this was necessary, because of the deficient education of the engineering students. This prohibited an integration of the technical instructions in the regular curriculum of the university. The instruction in the Dutch language was never fully accepted by the university community and the professors in the 'Duytsche Mathematique' had a lower status than the regular professors.[5]

The education of military engineers soon became less important, but the courses in surveying lasted for more than two centuries. Other universities also took up the training of surveyors. These courses prepared for an examination, necessary for becoming entitled to work as a surveyor, and were regulated by the provincial government. In most provinces a certificate of a university was sufficient to become qualified as a surveyor and it certainly added prestige to its holder. A second way to prepare for the examination existed in studying and working with an experienced surveyor. Still, the training of surveyors at the universities was the only form of a more formal technical education in Holland until the end of the eighteenth century. Some of these surveyors became involved in technical tasks pertaining to the most important area of administration in Holland, the control of the water system.

The struggle against the water had a large impact on Dutch history. The water was, due to geophysical properties of Holland – a large sea coast, many rivers and many polders under sea level – a constant threat. Various large floods, e.g. in 1726 and 1740/41, reminded the population of the potentially very dangerous situation. It is therefore quite natural that the history of civil engineering in Holland is mainly connected with the care for waterways and rivers, harbours, dikes and polders. In the 18th century people became more aware of the relations between the various parts of the water system and the term *Waterstaat*, literally the state of the water, became the common term, describing everything connected with the water system.[6] One could distinguish three main areas of the *Waterstaat*: 1. the care for the inner dikes, polders and draining; 2. the care for the maintenance of the sea-coast and 3. the management of the rivers. The care for the public works in these areas was given to independent public institutions like *waterschappen*, the polder-boards, another unique Dutch institution, which reflected the political constitution of Holland in the 17th and 18th century; the Netherlands were a federal republic of seven provinces with a large degree of autonomy for each of the provinces.

Practical-technical problems in the *Waterstaat* were handled by craftsmen with practical experience and knowledge. The public boards employed a large variety of technical personnel, like surveyors, supervisors, lock- and bridge-keepers and assistants who were all educated in practice with the exception of the already mentioned surveyors. Although for the higher positions often surveyors were asked, no general rules existed, due to the independence of the boards and the provinces. At the end of the 18th century things started to change gradually. There was a growing awareness of the necessity of a more integral approach of the *Waterstaat* and some efforts were made to centralize or at least coordinate the execution of public works, but this was because of the political situation still impossible.

After the failure of the attempts to centralize, however, the powerful Province of Holland decided in 1754 to create a new provincial function, the Inspector-General for the *Waterstaat*. The first man appointed in this position was Lulofs, a professor in mathematics and physics at Leyden University. Due to the work of Lulofs and his successors topics in the field of the *Waterstaat* were more general discussed. The technical aspects of the *Waterstaat* became indicated as the *Waterbouwkunde*, literally water-construction engineering.[7] In Holland *Waterbouwkunde* and civil engineering became almost synonyms and will be used in this paper interchangeably. The newly founded scientific societies also took part in the discussions on the *Waterstaat*, e.g. by offering prices for solutions of technical problems related to the *Waterstaat*, and the first books on this subject were published.[8]

The emergence of the *Waterbouwkunde* as a more or less coherent technical domain induced also the emergence of engineering courses. Civil engineering was introduced as part of the curriculum at the private schools of the 'Foundation of Renswoude', a kind of charitable institution for orphans, at Utrecht, The Hague and Delft. In the province of Zeeland a vocational school for engineers in the *Waterbouwkunde* was founded in 1791.[9] In the same period the first military schools were established. In the engineering corps of the army discussions on the training of officers had started already around 1750, but the education of future officers had a very low priority at the central government, defence policy being one of the few centralized domains. As a result of these discussions and following other countries three Artillery Schools were founded in 1789. However, these schools were not very successful and already before the French Revolution reached Holland, two of them were closed.[10]

3. Under French rule

The French Revolution was the starting point of some profound changes in the Netherlands. In 1795 the Prince of Orange was sent away by the patriotic movement. They proclaimed the Batavian Republic and a process of political centralization started. The patriotic movement disagreed however on the degree of centralization of political control. Various coup d'etats took place, each time swinging the pendulum from more centralisation to more federal influence and back. The general result however, was the emergence of a national, centralized state, and more specifically, the foundation of a central Department for the *Waterstaat*, called *Rijkswaterstaat* in 1798 and the formation of an Engineering Corps of the *Waterstaat* a few years later.[11] The training of military engineers was taken up at four schools, three of them more or less a continuation of the previous ones.

In 1806 Louis Napoleon, the brother of Napoleon Bonaparte, became King of Holland, and another turbulent period started. Louis Napoleon, having the French system in mind, wanted to reorganise the educational system in Holland completely. The existing military schools were reorganised to form one new school at Amersfoort. The competent director of the school at Zutphen, general Voet, became the first director. The king also ordered the foundation of a new military school following the example of the Imperial Military School at Fontainebleau. This school started in 1807 at Honselersdijk, but already in the same year the king decreed a fusion of the two schools into a new one, located at The Hague. At this new Royal Military School not only technical officers for the army should be trained but also civil engineers for the Engineering Corps of the *Waterstaat*.[12] The decree was however not carried out immediately.

The king also wanted to found an Ecole Polytechnique in his kingdom. A committee of the Royal Institute of Arts and Sciences, an imitation of the French Academy of Arts and Sciences, was installed in 1808 to investigate the possibilities and concluded that there were two ways of training engineers. The first one followed the French model, where general theoretical education at the Ecole Polytechnique was completely separated from the more specific vocational training at the other special schools. The second model was the combination of theoretical and practical training at one school, as was usual in Holland. Which system could be applied best in a country depended on factors like the number of inhabitants, geographical position and the career possibilities of its graduates. Because of the limited number of inhabitants in Holland, the relatively small need for educated engineers and the expected high costs, the committee advised the king very politely not to choose the French model. In this way another problem could be avoided. Some pupils could be seduced by the 'sweet taste of the Sciences'. Wanting to pursue their preferences, they would be less fit or even lost for the vocational training.

The report of the committee was sent in December 1808 by the minister of Interior Affairs to the king. To support the general conclusions he added that several capable civil engineers were educated at the private schools like those of the 'Foundation of Renswoude'. The king was not satisfied with the results, because he was convinced of the usefulness of a central school where the 'Physical and Mathematical Sciences' would be taught at the highest level. He ordered a new investigation, but the newly formed committee came to the same conclusions, although the arguments were a little bit different: the bad situation of the secondary education and also the potential loss of the 'esprit observateur', indispensable in practice, caused when pupils were too long subjected to only theoretical education was mentioned. To prevent a negative reaction by the king, however, a design for a Central School for Arts and Sciences was added. In a last vain attempt to save his ideals the king requested to investigate whether the combined military school couldn't serve as a Polytechnical School. Finally, the school at The Hague could start in november 1809. Again Voet was appointed director, but this school proved to be a very temporarily event.

In july 1810 Holland became part of the French Empire and all Dutch institutions were integrated in the French centralized system. The Royal Military School was closed and some of the pupils were placed at French military schools. The nine students in civil engineering were put at the disposal of the Director-general of *Rijkswaterstaat*. In the next year they were provisionally placed at the Ecole des Ponts et Chausses. These 'élèves' had a difficult time in France because they lacked the necessary preparation for a study at the Ecole Polytechnique. Also, financial circumstances were very precarious for the young students, but at least a few of them, especially I.P. Delprat, W.C. Brade, E. de Kruyff and F. Baud, would became important persons in the emerging engineering community in Holland.[13] Their study ended in 1813 after the first overthrow of Napoleon, and they returned to Holland. Already before or immediately after the final defeat of Napoleon in 1815 all Dutch officers and functionaries who had served in France returned to the Kingdom of the United Netherlands (now including Belgium and Luxembourg).

4. Civil engineering at the military schools

Already in December 1813 the provisional government asked the military governor of Amsterdam to investigate the possibilities of a new military school. He asked the experienced Voet to make a concept for a new school. In 1814 the Artillery and Engineering School was founded at Delft. At this school also a few civil engineers were educated. One of the first problems Voet had to solve was to find capable teachers.

As professor in the mathematical sciences Jacob de Gelder was appointed, a self- taught man, who previously had been lecturing at various private institutions and was generally regarded as one of the best mathematicians available.[14] Officers for the practical instructions were more difficult to get as most officers preferred an active career. So, the still very young former students at the Ecole des Ponts et Chausses Delprat, Brade and De Kruyff became assistants of the instructions in military engineering. C. Cox, an engineer of the department of *Rijkswaterstaat* became responsible for the instruction in civil engineering. Cox had been studying at Leyden, probably preparing for the surveyor examination.

In the first years there were no entrance requirements, which led to a large diversity in the level of the students. In the first few months their knowledge of mathematics was tested. The most advanced pupils were placed in the class of De Gelder, the others were instructed by Cox, Delprat and Brade. Voet argued also, that he wanted to construct a chemical laboratory in order to reach the same level as comparable institutions abroad. The military supervisors deemed this unnecessary, while studying a chemical handbook should be sufficient. Voet had to give up this project.

Another problem of a more temporary character was the exodus of students after the return of Napoleon from Elba in 1815. In the newly formed Kingdom of the United Netherlands under the dynamic rule of king William I, things settled finally. In the following years the school expanded. An entrance examination was introduced and various new teachers were appointed, for instance a lecturer in mathematics to assist De Gelder. Cox, De Kruyff and Brade left the school and went back to active service.[15] They were replaced by Baud who became responsible for the instruction in civil engineering.

The instructions in mathematics formed a very substantial part of the curriculum at the Artillery and Engineering School. Especially from the more traditional arms like the artillery, complaints were made on the 'strictly mathematical proofs' De Gelder demanded of his pupils. Voet asked De Gelder to simplify and shorten the mathematical courses. Voet complained that in the period before De Gelder, the courses in mathematics were given by officers, who knew what the students needed to know for their work afterwards.

But the approach of De Gelder, who insisted on mathematical rigour, was – according to Voet – fatal to the practical education of the students, which was proved by the experience with recently graduated engineers of the school. At the construction of new fortifications they had made a very bad impression and the commander complained to Voet on their lack of skills. The conduct of an assistant of De Gelder – "a mathematical fanatic" – was too much. A vehement controversy arose between Voet and De Gelder. De Gelder responded by sending a 'memorial report' directly to the king, in which he complained about the defective organisation of the school and proposed a new structure more in accordance with the French Ecole Polytechnique. He also wrote that too little attention was paid to mathematics. The furious Voet suspended the professor and a year later he was dismissed.[16]

The complaints about the graduates and the controversy on the place of mathematics in the curriculum led to a complete reorganisation of the school. The whole orientation became more practical. The instruction for the different arms and for the *Waterstaat* were separated. Only the first year was the same for all students. Besides, more time was spent on practical exercises. The students in the *Waterbouwkunde* e.g. had to volunteer after their third year at the construction of public works. In the discussions with the army commanders one issue was not solved. In all military schools the students were placed in military barracks.

Voet however, deemed the military system with its severe discipline not well suited for his students. The students at Delft stayed therefore with civilian families, much to the dislike of the generals. Eventually this led to the decision to close the Artillery and Engineering School at Delft and replace it by the Royal Military Academy at the castle of Breda, enlarged to accommodate all students. In 1828 the school at Delft was closed. Voet stayed at Delft. At the end of his career he acknowledged that mathematics had been too prominent at the military school. It should have been an auxiliary science and not the principal science in the curriculum. In the 14 years of its existence 42 students started studying civil engineering, 24 of them graduated.

The entrance requirements in Breda were increased, compared with the Delft School. The candidates should have a thorough command of arithmatics and should at least know the basic principles of algebra and geometry (triangles, polygons and circles). The majority of the students and nine teachers of the Delft School moved to Breda. Baud however, left the school. The instructions at the school became already in the second year severely interrupted by the secession of Belgium in 1830. Most of the officers and many students were drafted. The actual fighting lasted only some ten days, but the army was mobilized for many years. In this period the school, at least what was left of it, was temporarily moved to the Institute for the Marine at Medemblik, because Breda was too near to the new border. In 1836 the school went back to Breda, but only after the treaty with Belgium in 1839 things became normalized again.[17]

Delprat, one of the teachers who moved to Breda, became responsible for the instructions in mathematics and the natural sciences and in 1836 he was promoted second commander of the Royal Military Academy. He became one of the outstanding and most influential persons in the Dutch engineering community.[18] Already in 1821 he published a book on the retaining wall problem. This problem was important in a military context. Military engineers needed to know the minimum strength of retaining walls. The usual way to solve the problem was to overdimension. The first theory on this subject was formulated by Coulomb in the 18th century. Delprat treated in his book the theory of Coulomb and his successors, but also criticised these and proposed some improvements. In the last years of his long life he published a complete revision of this book.[19] He also wrote a book on the trajectories of projectiles (1826) and on the resistance of iron beams (1832). In all his books and articles he proved that he was well acquainted with the most recent literature, but he did not merely translate foreign texts but always made a contribution of his own, e.g. by extending the application of new theories or by simplifying formulas. Because of his scientific contributions, he was one of the few engineers to be elected a member of the Royal Dutch Academy of Sciences.

The most important contribution of Delprat concerned the education of military and civil engineers. Already at Delft, one of the complaints was that there were no engineering textbooks available in Dutch. Therefore foreign books had to be used and this hampered the instruction. So, Delprat decided to write some textbooks himself. In 1840 and 1842 his books on mechanics and mechanical engineering were published, which dealt with statics, dynamics, hydrostatics and hydrodynamics. He always started with theoretical considerations but also gave many examples of applications of the theories. These books were reprinted several times. He also encouraged his colleagues to write textbooks. He himself revised the texts, used by Baud for the instruction on several subjects on civil engineering (theoretical and practical consideration on the construction of locks and sluices). J. Badon Ghijben wrote several mathematical textbooks on advanced geometry and differential calculus, G.A. van Kerkwijk wrote a book on geodesy and D.J. Storm Buysing wrote a two volume handbook on civil engineering. Before publishing, Delprat corrected and improved all these texts. Most of these books were reprinted once or twice and although meant for the instruction at the Royal Military Academy, they were extensively used, also on the new school for civil engineers at Delft, which started in 1843.

As was clear from the start, the education of civil engineers held only a very marginal position at the Royal Military Academy. Of the maximum number of 300 students only 4 places were reserved for the *Waterstaat*. In total 19 civil engineers graduated at Breda between 1828 and 1845. Nevertheless, both the military school at Delft and at Breda played a very important role in the formation of the engineering profession in Holland. After 1816, almost all new members of the Engineering Corps of the *Waterstaat* were graduates of the military schools.[20] In this way, a formal education became a necessary condition for entering the engineering profession and this marked the transition to a school culture. In the first half of the 19th century, the size of the Engineering Corps was limited to some fifty members. Because the civil engineering education was almost exclusively directed at the Corps, only a few graduates could be admitted every year. Their social position until 1850 is described as somewhat marginal, compared with the military Corpses and they only had partial control over the technical domain of the *Waterstaat*.[21] Around 1850 things started to change rapidly. After the foundation of a new school the number of civil engineers also started to rise sharply.

5. The long road to a Polytechnical School

The ideal of King Louis Bonaparte to establish a Polytechnical School in Holland was not completely forgotten after the French period, but it lasted 35 years before it was accomplished to a certain degree. The new King William I was a dynamic person, who stayed in England during the French period. After returning he was determined to follow the English example to industrialize his country. He improved the infrastructure by building a network of canals and he tried to stimulate the creation of a modern industry by opening up the colonial markets. Especially the more industrially developed southern part reaped the benefits of this policy. William I also intervened in the educational system by reorganizing the universities in 1815. In the southern part three universities were established, at Gent, Leuven and Liège. In the northern part also three universities remained, at Leyden, Groningen and Utrecht. In the new system no place was left for the 'Duytsche Mathematique' probably because the king had other ideas on the education of surveyors and engineers. It took ten years before the king issued a royal decree: all six universities should provide lectures on the application of chemistry and mechanics to the useful arts. As these courses were specially meant for craftsmen, they should be given in the vernacular.

The universities reacted in different ways upon this rather vague instruction. At the University of Groningen, for example, G.J. Verdam was appointed lecturer on applied mechanics. Dreaming of a military career, Verdam had studied at the Artillery and Engineering School at Delft, but after the dismissal of De Gelder in 1819, he had left the school and moved to Leyden, where he studied mathematics and natural sciences at the university. After writing a dissertation in 1825, he was sent by the government to the Cockerill works at Lige to obtain some practical experience. After visiting some textile factories, he started to lecture at Groningen. However, his courses were no success. The appreciation of his lessons at the university was very low and very few persons – no regular students – attended his lessons. Disappointed, he left Groningen already in 1828 and founded a secondary school at The Hague; he was also appointed inspector of steam boilers and engines. Between 1829 and 1837 he published an extensive work (six volumes) on the principles of applied mechanics, especially on steam engines. This work, including an introduction to arithmetics and geometry, was the only original Dutch book on steam technology to be published in the 19th century. It was translated into German by Ch.H. Schmidt.[22]

The example of Groningen is illustrative of the experiences with the special courses for industry. Although a course on chemistry for a general public at Leyden was continued for more than thirty years, the overall results were very disappointing and within a few years all courses disappeared. One of the reasons for this failure was the condescending reaction to the technical courses in the vernacular by the university community.[23] In the same period the king asked a special advisory committee to evaluate the system of higher education. The position of the technical education was mentioned explicitly in the instructions of the committee: was it necessary to give courses at the universities for training civil servants and for promoting commerce, agriculture and industry? If so, would it not be better to give each university a field of specialisation in agreement with the local specialisations? Another possibility, the committee had to evaluate, was the transformation of one or two of the universities into two polytechnical schools.

The reaction of the committee on these proposals was very negative.[24] The first task of the universities should be the scientific education of the 'learned classes'. More practically oriented lessons could never be fully integrated in this system. Besides, the foundation of polytechnical schools was not necessary, because the few civil engineers needed, were already trained at the Royal Military Academy. The small number of mechanical engineers, needed for industry, could be better encouraged to study abroad. If nevertheless the government wanted to found a polytechnical school, a necessary condition would be the improvement of the secondary education.

The king was not convinced at all by the recommendations of the committee. Immediately after the publication of the final report he announced the establishment of a Royal School of Arts and Sciences at Brussels. This plan was probably the result of discussions between the king and the capable engineer A. Lipkens. Lipkens had studied as an extra-ordinary student at the Ecole Polytechnique and had been for almost ten years a teacher in surveying at various French military schools.[25] After returning to Holland he first became engineer at the land registration office at Luxembourg, where he had to train his own assistants, because they lacked any kind of education in surveying. Later he was appointed special advisor on patents for the ministry of interior affairs. As could be expected, Lipkens was convinced of the necessity of a polytechnical school and apparently he succeeded in convincing the king. The king asked Lipkens to become the first director of the school in Brussels, but all these plans were prevented by the secession of Belgium. The new government of Belgium asked Lipkens in 1832 to carry out the plan, but probably because of his personal loyalty to the king, he refused. He did not give up his idea of establishing a polytechnical school in Holland, but he had to suspend this plan for some time.[26]

Around 1840 Lipkens saw new possibilities. The conflict with Belgium was finally settled and the weary and disappointed King William I was replaced by his son William II. Lipkens was well acquainted with the new king. In January 1842 the King announced the foundation of the Royal Academy at Delft for the training of civilian engineers for public service, for industry and for commerce.

Lipkens would become the first director. At that moment only a general plan existed and a lot remained to be done. Due to the very precarious financial situation of the state, no subsidies would be given. With the exception of a starting grant the engineering school had to be financially independent. The number of students and therefore the amount of lecture fees was considerably increased by adding a training programme for functionaries in the Dutch East Indies. Lipkens, though, had initially not been very happy with these students, because they didn't belong to a real engineering school.[27]

The Royal Academy was a very peculiar institution. It was a kind of private enterprise of the king, who appointed his son as patron of the school and delegated the supervision to the Minister of Home Affairs, who was represented at the school by the director. At the school students were trained in civil engineering, mining, mechanical engineering, architecture, shipbuilding and commerce, along with civil servants for the Dutch East Indies (first class and second class) and functionaries for the Treasury. In this paper I will deal only with the training of engineers. Between 1842 and the end of the school in 1863 no student graduated for commerce, 17 in mining – all going to work in the East Indies – 5 in naval engineering, 2 in mechanical and chemical engineering. 183 students graduated in civil engineering and almost twice this number started this study. Without exaggeration one can describe the Royal Academy as a school for civil engineering. Already at an early stage the best students were selected for the *Rijkswaterstaat*. After a final examination they were admitted directly to the Engineering Corps of *Rijkswaterstaat*. The number of students selected, depended on the demands of *Rijkswaterstaat*. The other students followed the same programme. After the final examination they got a diploma.

The first two years of the four year curriculum consisted mainly of mathematics, some physics and chemistry. Professor in mathematics, assisted by two teachers, was dr. R. Lobatto, who had a very good reputation. When the number of students rose in the fifties, more teachers in mathematics were added. Compared with mathematics the lessons in civil engineering took only a small part of the curriculum. In the third year the subject of the civil engineering course was the description of the main public works. In the last year, a description of the construction of public works in water and the materials used for this type of constructions were on the programme. The students also learned to design locks, bridges and buildings and, finally, to calculate the costs and to make the specifications for the execution of the plans. The training was completed by a description of useful tools and instruments and knowledge of the hydrography of Holland. There was only one teacher in civil engineering, M. Beijerinck, chief engineer of *Rijkswaterstaat* gave the lessons in civil engineering. After Beijerincks death in 1847, he was replaced by Storm Buysing, teacher at Royal Military Academy.

Both Beijerinck and Storm Buysing were graduates of the former Artillery and Engineering School. Until 1859 Storm Buysing, who soon became professor, was the only one teaching civil engineering at Delft. Because of the appointment of another teacher in that year, the lessons could be extended somewhat. The drawing and designing of bridges, roads and railways and the various constructions connected with them were assigned to the new teacher. Besides, they learned to calculate the amounts of materials needed. The calculation of the strength of the constructions proved to be too difficult for the students, who lacked the necessary knowledge of statics.[28]

Although we don't know exactly what the students learned, we get a good impression from the textbook on *Waterbouwkunde* for the Royal Military Academy by Storm Buysing, published in two volumes in 1844/1845, reprinted ten years later. In volume one roads, railways, the various types and parts of bridges and the construction of dikes are treated. In volume two the main subjects are the rivers, navigation, the construction of locks and sluices, ports and draining. Large parts of these books are purely descriptive, for example the construction of dikes and the materials used for this kinds of works. The knowledge contained in these parts was based on practical experience. Only elementary mathematics is used to describe these works, e.g. describing the angles. In fact, only the sections on railways and railway bridges contain some theoretical considerations, but in these sections also very few formulas are used. Some exceptions are a formula by De Pambour giving the power of steam locomotives and one for calculating the maximum load of the iron beams (without losing its elasticity). This last formula was taken from the article by Delprat on iron beams. In these cases both French, English and German literature is quoted.

Summarizing, one can state that the specific technical knowledge civil engineering students around 1850 learned was to a high degree practical and empirical. For this purpose they didn't need the extensive mathematical training. Because often references were made by civil engineers to books like Storm Buysing's one, one can deduce that the graduates also needed little theoretical knowledge in practice. The only exception was the field of railway engineering, but this new technological domain was in Holland only in its initial stage at that time.

The position of the Royal Academy stayed precarious for some time. In 1848 some members of parliament proposed to close the school, which was deemed to be superfluous and too expensive. The last argument was a little bit strange, because the Academy was financially independent, but maybe some MP's were not very happy with the status of the school, which was a kind of private affair of the king, governed by royal decrees. The first argument was certainly understandable. Only a few students every year were engaged by the Engineering Corps and in the beginning there were no other career perspectives for the graduates. Only after the decision of the Government to take over the construction of the railway system in 1860 these engineers were offered new possibilities.In order to support the interests of the graduates of the Delft school in these difficult times in 1853 the Association of Engineers from Delft was founded. This organization resembled the existing school associations in Germany and Belgium. Later it focused more on the promotion of the social interests of the civil engineers. The Royal Institute of Engineers on the other hand, founded in 1847, was more like a scientific society, directed by the most prominent engineers. The director of the Royal Academy was one of the founders of both organizations.

The Academy was saved temporarily by the resignation of the Government. In the end of the fifties a recurrent theme in the engineering education in Holland surfaced again. Many complaints were made on the lack of discipline at the school and the presumed scandalous behaviour of its students. After the appointment of a retired officer as director of the school, a conflict broke out between the new director on one side and the students and some of the staff members on the other. This resulted in a suspension of the lessons for two months at the end of 1861. The new law on secondary education in 1863 led to the definitive end of the Royal Academy. The Academy was replaced by the Polytechnical School.

6. The Polytechnical School

The law of 1863 regulated for the first time the system of secondary education. The existing gymnasia prepared for the universities. Other secondary schools were mainly private schools, some offering a general curriculum, others with a more technically oriented curriculum. As a result of this law several secondary schools of a new type, the *Hogere Burger School (HBS)*, to be compared with the German Oberrealschule, were established and within a few decades every small town in Holland possessed a *HBS*. Mathematics, physics and chemistry formed a major part of the curriculum. The *HBS* prepared explicitly, but not exclusively for the new Polytechnical School (PS) at Delft. Within a few years the problem of the entrance conditions was completely solved. As a result the Polytechnical School also belonged to the secondary education. For the Government this was the logical outcome of the two separate paths to higher education: one from gymnasium to the university and the other from the *HBS* to the PS or another vocational school. For the engineering community this meant however an inferior position of their school compared to the universities. Until the elevation of the PS to university level in 1905 this remained one of the main issues of the Dutch engineering community.

In some aspects the PS was a continuation of the Royal Academy: it was located at the same place and a considerable part of the staff was employed at the new school. But there were also many differences. The PS was exclusively an engineering school. The education of functionaries for the Dutch East Indies and for the Treasury disappeared. The PS also became a true polytechnical school, offering the opportunity to study civil engineering, architectural engineering, naval engineering, mechanical engineering, mining and industrial or chemical engineering, called technology. All studies were four years, with the exception of the study of technology, taking only three years. The first two years of all directions were more or less the same, including mathematics – more specifically courses in plane and solid geometry, analytic geometry, differential and integral calculus, vector analysis and probability – theoretical mechanics and applied mechanics (the mathematical analysis of static and dynamic systems, calculations of strength of materials and structure) physics and, voluntarily except for the technologists, chemistry. After these two years the students had to pass an examination. The last two years, also concluding with a final examination, were more discipline oriented.

In spite of the diversity of disciplines, civil engineering stayed by far the most important field until the eighties . The civil engineers also dominated the professional organizations. In fact, the civil engineers seized the opportunities offered by the construction of a national railway system after 1860 and took complete control of this new technological domain. This was reflected by the rapidly increasing number of the students in civil engineering: 20 in 1850, 225 in 1878, dropping to only 70 in 1889 after the completion of the railway network. In that year civil engineering was still the largest department at the PS, although mechanical engineering was for the first time starting to challenge this dominant position. In this paragraph I will confine myself to civil engineering.

The subjects of the final examinations in civil engineering were theoretical and applied mechanics (applied to tools and engines), the *Waterbouwkunde*, civil architecture, drawing, design of specifications, principles of geodesy and surveying and finally law. The descriptions of the subjects of *Waterbouwkunde* reads in fact as a table of contents of Storm Buysing's book. These courses were taught by Lebret, engineer of the *Waterstaat*, from 1869 until 1884 by E. Steuerwald and afterwards by J.M. Telders, who had worked on the construction of the railways. Railway engineering, especially the construction of railway bridges, became naturally a prominent topic in the instructions. Already in 1866 the courses on roads and bridges became a separate part of the curriculum. For this purpose, N.H. Henket was appointed professor in the *Waterbouwkunde*.[29] After this adaptation, the programme only changed slightly until the nineties.

In 1877 three professors at the PS, Henket, Steuerwald and dr. Ch. M. Schols

– professor in surveying, levelling and geodesy and graduated in civil engineering at Delft – started an ambitious project. They wanted to give a complete description of the state of the art of the *Waterbouwkunde* in the Netherlands. The whole field was divided in four parts, the first ranging from retaining walls to polders and drainage systems, the second on the rivers, the third on bridges and the fourth on roads and railways. Although between 1878 and 1899 several volumes appeared, (at least fifteen), the project was never completed.[30] But these handbooks give us the possibility to get an impression of the progress in civil engineering in Holland between 1850 and 1890. I will concentrate on soil mechanics and railway bridges.

In 1885 the first volume of the first part on foundations, camp-shedding and retaining walls was published. The first chapters, by Steuerwald, gave a general description. The main part, written by Schols, is much more interesting. After 1850 many prominent engineers had turned to the problem of the retaining wall. The pressure of the soil against a wall was the first subject of Schols. In a note he wrote, that the more rational theory of the soil pressure, based on the internal equilibrium of the mass of soil, developed by Scheffler, Rankine, Winkler, Levy, Considre and others, still didn't yield results, which were generally usable. The differential equations, given by these new theories, could only be solved for one special case, in which the plane of sliding of the soil is flat. Schols concluded therefore that one could start from this situation. This assumption resulted in values, that were slightly too large, but this was of course no problem. After this remark he gave an extensive mathematical exposition of the solution of the problem. Despite all the mathematical thoroughness of this exposition and of others, e.g. by Delprat, there is no evidence that in the 19th century this theory was really used in practice in Holland by the civil engineers.[31]

This can also be concluded from comments by Schols in his next chapters on camp-shedding. This type of construction, for example used in rivers and canals, was much more common in the Netherlands. In the opening remarks Schols wrote, that he would try to calculate the dimensions of the camp-shedding, but this had been a difficult task, because he had found no reference at all to this subject in literature. He was quite pleased with his results, which for the most common cases deviated not too much from the usual practice, in all other situations he advised to rely on experience. It is not probable that these calculations had a large impact on the work of civil engineers in this field. The development of soil mechanics really started after 1925 and it seems that the most important function of all the attention given to the subject of soil pressure was to emphasize the scientific character of the civil engineering education. The students at the PS, at least, were well trained in this problem.[32]

The importance of the railways for the civil engineering community was clearly reflected in the five volumes on railway bridges. The last volume, published in 1896, was completely dedicated to the calculation of bridges, which contrasted sharply with the limited elaboration of this subject by Storm Buysing. The design and construction of the bridge over the Lek at Kuilenburg, between 1862 and 1868, had been the turning point in this field of engineering in Holland. With the expansion of the railway system the problem of crossing the large rivers had to be solved. Two civil engineers, Van Diessen and Schneitter, were sent by the minister to Germany to investigate the new railway bridges. In addition to a report of their study tour, they wrote an extensive theoretical paper on bridge building. It started with a survey of the new theories of Schwedler and Cullman on truss bridges. The second part was on the usual and better known methods for the calculation of strength and the deflection of beams. The other parts of the paper contained data on the strength of materials and a survey of the various possibilities of connecting the parts of the bridge. Their work convinced the engineers that a truss bridge with a very large span crossing the summer bed of the river was feasible. The main span of the bridge at Kuilenburg with a length of 150 meters was, for a few years, the largest single span in the world. Not only in Holland, but also abroad, this was regarded as a remarkable achievement, a miracle of technical ingenuity, and this bridge served for some decades as the exemplar for railway bridges in Holland. The methods and theories, used by Van Diessen and Schneitter, expanded with the theories and methods of engineers like Ritter and Mohr were soon introduced in the curriculum at Delft. From about 1870 a special course on construction theory was added. Designing and calculating a bridge became one of the favourite topics at the final examinations.[33]

Although this gives the impression that after 1860 the designing of bridges rested on a firm scientific basis, it is necessary to be careful. In a review on the development of railway bridges a civil engineer wrote that it was impossible to take into account all the factors and circumstances, influencing the behaviour and strength of the bridge. One had to do with an approximate, already very complicated calculation. When the engineers tested the bridge with certain maximum loads, as was prescribed by the government, they measured the deflection of the beams. When there was no match between the measurements and the calculations – as was usual – the engineers made no comments on the strength of the bridge, but considered the measurements an excellent test of the theory![34] Because of these problems with the analytical method (= calculating the forces) the graphical methods, developed by Ritter and expanded by others, were much preferred by civil engineers. The analysis of structures by graphical methods was called the graphostatica and the diagrams it produced, the Cremona diagram. In 1879 this approach was applied to bridges by the Civil engineer A.B. Reintjes.[35] For nearly a century, this became a regular part of all courses in civil and architectural engineering.[36] In this period graphical solutions to technical problems became very popular in various fields, e.g. the solution to the retaining wall problem, but also for the designing of compound steam engines.

7. Theory and Practice

We have seen that the engineering education in the Netherlands started with the school at the University of Leyden. The second important origin was the emergence of the *Waterstaat* and its technical domain, the *Waterbouwkunde*. The need to train technicians in this field was a stimulus for the engineering education, although the success was limited, due to the political situation. The breakthrough came during the French period, firstly, because of the political centralization and the emergence of a central state Department for the *Waterstaat* and its Engineering Corps and secondly, because of the profound influence of the French educational system. The French influence becomes apparent in the content of the curriculum with its emphasis on mathematics and also in the engineers, becoming the first educators. The engineers, who had studied in France, were the most important teachers at the engineering schools.

The French model of the Ecole Polytechnique was not adopted, because it did not fit in the Dutch educational system. In the Dutch system theory (mathematics) and practice had to be combined at one school, but there was still a problem of integration. In fact, almost no integration took place during the main part of the 19th century. In spite of the lack of integration between theory and practice, the training of civil engineers – after the first problems at the Artillery and Engineering School – proved to be adequate for the work the engineers had to do. The general attitude of the Dutch civil engineers was a very practical one. They relied on their experience and, although they were well aware of the theoretical developments abroad, they were not tempted by a theoretical approach to the problems encountered in their work.

So, the extensive mathematical training was not a necessary requirement for the work of the engineers. The retaining wall problem is an illustrative example of the use of theory: the students had to prove their ability by this problem, but they didn't use it afterwards. In this regard, the extensive mathematical training could only serve as a way to select future engineers. This started to change gradually after 1860 with the construction of the railway bridges, becoming the most important topic of the community of civil engineers. The combination of the control of this domain by the civil engineers and the changes in the knowledge base of the engineers made a more formal training a necessary requirement for the first time. For these new developments the Dutch engineers were, however, well prepared.

The requirement for a more formal study to enter the Engineering Corps after 1816 marked the transition to a school culture. This 'military segment' within the Engineering Corps became the main advocate of a formal training of engineers.[37] Within the Corps they competed with their superiors, who were only trained in practice. The argument of the (lack of) education was used several times in comments on the organization and the deplorable position of the Corps. Around 1850, the military segment succeeded in taking over the control over the Corps and soon also over the technical domain of the *Waterstaat* and also of the railways, which resulted in a rise of social status and a growing influence in society. From this time, they could defend the interests of the academically trained engineers themselves.

In the first half of the 19th century this situation was quite different. With the exception of some modernizers, like King Louis Bonaparte and King William I, there was little support for an engineering school. The resistance or at the best a lack of interest in technical training at the universities was mentioned. But also in general, in Dutch society a commercial attitude was predominant. To put it in a popular way, every proposal was subjected to a cost-benefit analysis. The result of this analysis was always a lot of costs and almost no benefits. This was, after all, not a bad evaluation. So, one can conclude, that the failures to found a Polytechnical School, and, as a consequence, the education of civil engineers at military schools, proved to be a blessing in disguise. At these schools, the engineers were protected against the pressure and influence of a hostile society and they could quietly build their own empire. The price they had to pay, the military discipline, uniforms and a hierarchical organization, was a small one, compared with the advantages.

Notes

1. Peter Lundgreen, 'Engineering education in Europe and the USA, 1750-1930: the rise to dominance of school and the engineering professions' *Annals of Science* 47(1990) 33-75
2. On the history of the Dutch engineering profession and schools has been written by: H.W. Lintsen, *Ingenieurs in Nederland in de negentiende eeuw* (The Hague, 1980); C. Disco, *Made in Delft* (Leyden 1990) and H. Baudet, *De lange weg naar de TU Delft..*
3. S.Stevin, *De sterctenbouwing* (Leyden 1594)
4. The subjects of mathematics were adding, subtracting, multiplying, dividing an extraction of roots and the calculation of the area of squares and triangles.
5. P. J. van Winter, *Hoger beroepsonderwijs avant la lettre* (Amsterdam 1988)

6. S.J. Fockema Andreae, 'Centraal Waterstaatsbeheer', in: *Publicaties van het genootschap van napoleontische studiën*, afl. 1 1951, p.30.
7. Waterbouwkunde is commonly but unjustly translated as hydralic engineering. The Waterbouwkunde also includes the various methods of constructing and protecting dikes and the construction of bridges.
8. The most important scientific societies were the Dutch Society of Sciences (1752), the Batavian Society of experimental Philosophy (Rotterdam 1769), the Zeelandish Society of Arts and Sciences (1769). One of the first more general books was the '*Inleiding tot de waterbouwkunde, bevattende de voormaamste gronden der beweeg-, waterweeg- and waterloopkunde*' by A.Van Bemmelen (Leyden 1793)
9. Fockema Andreae, 'Central Waterstaatsbeheer' p.32
10. J. A. M. M. Janssen, *Op weg naar Breda* (The Hague 1989) p.999-121.
11. H. W. Lintsen, *Ingenieurs in Nederland in de negentiende eeuw*, p.45-59
12. Janssen, *Op wegg naar Breda*, 233-253
13. On this point, I disagree with Janssen, who describes the journey of the Dutch students to France as a complete failure.
14. Jacob de Gelder (1765-1848) had been lecturer in mathematics at the Society Diligientia and at the institution, educating 'Les Pages du Roi'. After his dismissal he became an extra-ordinary professor in mathematics at the university of Leyden. De Gelder published several books on mathematics, e.g. *Beginselen der cifferkunst*, Rotterdam 1793, *Wiskundige lessen* (The Hague 1809), *Beginselen der meetkunst* (2 vol.) (The Hague 1810), *Meetkundige analysis* (2 vol.) (The Hague 1811-1813) and *Beginselen der differentiaal, integraalen variatierekening* (The Hague 1823). At the Royal Engineering School he used several of his books and also a parts of his books on calculus.
15. Brade left the army in 1819 to become one of the first civil engineers, working as an independent advisor. He became one of the first Dutch experts on the new railways. He was involved with the plans for the Dutch railway between Haarlem and Amsterdam and he was for a short period engineer-director of the company, exploiting this line. Between 1827 and 1834 he wrote four volume '*Theoretisch en praktisch bouwkundig handboek*', reprinted 1842. Especially the fourth volume on railways was influential.
16. Janssen, Op weg naar Breda, 307-317
17. The Royal Military Academy at Breda still exists.
18. An extensive biography of Delprat (1793-1880) was published in the Journal of the Royal Institute of Engineers in 1883. Delprat was between 1860 and 1866 president of the Institute. Although in his time, he was recognised as one of the outstanding engineers of his time, in recent literature on the history of the engineering profession by Lintsen (1980) and Disco (1990) he hardly is mentioned.
19. I.P. Delprat, *Verhandelingen over de zijdelingsche drukking der aarde tegen bekleedingsmuren en beschoeijingen*, Delft 1821 and I.P. Delprat, *Over de drukking van aarde tegen bekleedingsmuren, benevens ontwikkeling der voornaamste formulen voor de afmetingen deze muren. Ten gebruike der Koninklijke Militaire Akademie*, (On the pressure of soil against retaining walls, combined with a exposition of the most important formulas for the dimensions of these walls, intended use at the Royal Military Academy), Breda 1860, reprinted 1872.
20. H.W. Lintsen, *Ingenieurs in Nederland in de 19e eeuw*, p.73-78; In fact, only two exceptions on this rule are known.
21. Lintsen, *Ingeniuers in Nederland in de 19e eeuw*, chapter 6
22. Schmidt was the editor of a series of book on technological subjects, published by a society of artists and craftsmen ("Professionisten") at Weimar.

G. Verbong

23. At Gent, the Royal Decree resulted in the foundation of a Special School for engineering. This school was closely connected to the faculty of Sciences of the university and remained so afterwards. The main motive for organising the engineering education in this way was to reduce the costs. The history of this school is described by A.M. Simon van der Meersch in *150 Jaar ingenieursopleiding aan de rijksuniversiteit Gent (1835-1985)*
24. After a year of deliberations, the comittee, presided by W.F. baron Roëll, Minister of State, published a report of 400 pages. The part, dealing with technical education, contained only twelve pages. Also, there was no unanimity on this subject. Here only the opinion of the majority is given. H. Baudet, *De lange weg naar de TU Delft* (Delft 1992) 131-134.
25. Lipkens was born at Maastricht. Because Maastricht was annexed in the French empire already a long time before the rest of Holland, it was possible for him to take part at the entrance examinations for the Ecole Polytechnique at Gent. For obscure reasons, he arrived too late for these examinations and could therefor only be admitted as an extra-ordinary student.
26. Baudet, *De lange weg naar de TU Delft*, 152.
27. Baudet, *De lange weg naar de TU Delft*, p.171.
28. *Gedenkschrift van de Koninklijke Akademie en de Polytechnische School (1842-1905)* (Delft 1905) p.175. This Memorial contains a wealth of information on the education of engineers.
29. Henket (1829-1904) passed the surveyor examination in 1848, he worked at various projects of the *Rijkswaterstaat*, also a few years on Java. Due to his extensive practical experience, Henket was in 1866 appointed as a professor at the PS and he was between 1895 and 1897 director of the school. In 1899, after 34 years of teaching, he retired.
30. *Gedenkboek van de Koninklijke Akademie en de Polytechnische School* (Delft 1905) 122-123.
31. N.J. Cuperus, De ontwikkeling van de grondmechanica in Nederland tot circa 1940, *Jaarboek voor de Geschiedenis van Bedrijf en Techniek*, 6 (1989), p.30-53. Also, no reference is found to this theory, or any other theoretical considerations, in the reports of the preparation and execution of the large public works, e.g. the construction of the Northsea Canal.
32. In 1888 e.g. the students were asked at the final examination to investigate the stability of a retaining wall. They had to use graphical methods to solve this problem. The analytical and graphical analysis of constructions, like retaining walls, bacame a seperate course, probably already around 1870. *Examenvragen C2 voor Civiel- en Bouwkundig Ingenieur vanaf 1883* (Amsterdam).
33. G. Verbong, 'Civil engineers and the construction of railway bridges in Holland', paper for the Conference "Technological Development and Sciences in the 19th and 20th centuries", University of Technology Eindhoven, 6-9. nov. 1990
34. G.Verbong, 'Civil engineers and the construction of railway bridges in Holland'.
35. A. B. Reintjes, *Graphische behandeling der verschillende bruggen-systemen door middel van het influentie-polygoon* (The Hague 1879)
36. Only recently the Cremona diagram is increasingly replaced by computer programs.
37. Lintsen, *Ingenieurs in Nederland in de negentiende eeuw.*

Danish Polytechnical Education between Handicraft and Science

Michael F. Wagner

Introduction

A Danish polytechnic Academy of higher Education (*Polytechnisk Læreanstalt*) opened in Copenhagen in 1829. This brand new institution was seen as an integral part of the University of Copenhagen. The Academy would provide the candidates with a string of courses in the natural sciences at top level. While the courses in mechanics, technology and the workshop practice were preconcieved and later conducted in such a poor manner, that they played a minor part in the education of polytechnical candidates for the next many years to come.

The paradox of this situation is that the original address to the King proposing a polytechnical school, didn't aim at an academic institution at all. The primary intention had been to create a school that would educate the artisan and the craftsman after the 'German pattern' (Gewerbeschule). This strategy to heighten and improve the crafts was radically changed, soon after the King passed the proposal on to the Board of Directors of the University. This Board appointed a Committee of University-Professors to evaluate the original proposal, and then come up with an alternative proposal for a scientific polytechnical institution modelled after the 'French pattern'(Écolè polytechnique). The new proposal was more or less identical with the final plan for a polytechnical Institution enacted by the king in 1829.

Three political fractions and their policies will be identified in the paper, as will the string of extra-curricular activities undertaken by the members of these fractions. These activities include a society for the popular education in natural science, publication of a polytechnical magazine and textbooks.

As a conclusion the impact of *Polytechnisk Læreanstalt* on Danish industry and society during the 19th century will be assessed in different ways.

The Danish culture of technology

First I will present the Danish culture of technology in the 19th century. Technology is here taken as a concept of knowledge or software, rather than as a concept of machinery or hardware. The culture of knowledge was created by people in the scientific milieu around *Polytechnisk Læreanstalt*, the first technical university, which started to educate polytechnical candidates in November 1829.

The cultural concept of technology is constituted by the following three primary sources, each adding to the general technological culture in Denmark:

a) Technological institutions, mainly *Polytechnisk Læreanstalt*. From 1858 also Den Kgl. Veterinær og Landbohøjskole, (the Royal Agricultural University), that soon would be spoonfeeding Danish dairy industry.
b) Technological societies and public organizations dealing with polytechnics in one way or another.
c) Polytechnical literature, educational books and technical magazines.

The primary focus will be on the preparation and formation of the first Danish polytechnical university. *Polytechnisk Læreanstalt* soon became a driving force in a broader cultural movement and achieved a dominant technological position in Danish society. It contained a large amount of resources, compared to any other technological institution or sector in society.

Economic and intellectual resources were employed to educate civil servants. Not, as could be expected, to improve the handicrafts and develop an industrial production. At least, this was a secondary purpose of the education. What took place was the application of natural sciences to a polytechnical institution at the University of Copenhagen. The academic institution was then designed to produce government officials. And it had very little to do with artisans and handicraft or trade and industry at all. Anyway, a serious demand for engineers in production did not occur at any significant level until the 1880s and the institutions inability to meet this demand gave room for a lot of criticism from industries.[1]

The following discussion will touch upon these three sectors in the early development of the institution. But they will not be seen as independent factors in the analysis.

The early polytechnical tradition 1798-1827
A first book of technology

The first point of departure is a sketch of polytechnics before it was institutionalized. A remarkable occurrence was the publication of German professor Johann Beckmann's, Anleitung zur Technologie, in Danish in 1798.[2]

The reason for this publication is somewhat obscure. Beckmann's work aimed at the education of government officials. German cameralism[3] had been a part of the curriculum at the Sorø Academy in the 1760s and 1770s. At the turn of the century the school of cameralism was at a very low peak at the University. Cameralism was first to blossom again after 1820, when the Sorø Academy was reorganized. The reorganization was done by the Board of Directors of the University and the Learned Schools[4]. Soon after the Board became a decisive partner in the foundation of *Polytechnisk Læreanstalt*.

Beckmann's book probably had little impact on the development of Danish polytechnics. It appeared in a technological vacuum, and external affairs of state were soon to place a heavy stress on the Danish society. The Napoleonic Wars and the British naval attack on Copenhagen in 1801 and again in 1807 placed a severe stress on Danish economy. The following recession resulted in a fiscal collapse of the state finances in 1813. There was little room for growing expenditure in the decades that followed.

The translation of Beckmann's book into Danish may rather be illustrative of the total abscence of a polytechnical tradition or science. Cameralism and polytechnics are academic ambitions to systematize and describe technology. Cameralism as a purely descriptive economic science of production, while polytechnics is an experimental natural science of artefacts. Denmark possed neither of these academic traditions.

The translator's foreword is enlightening in this connection. He states his problems with the creation of a terminology to match the German technical expressions. This 'technology transfer' had been difficult because the Danes lacked a language of technology. Such a statement could be seen merely as an expression of academic snobbism. It can illustrate the social distance and lack of understanding between the refined academic writer and the crude artisan in the workshop. They did not speak the same kind of language and simply could not understand each other or communicate to a very high degree.

This may certainly have been the general case, with a few and very remarkable exceptions. However, it's not the problem in question here. The language gap between German concepts of technology and the Danish language demonstrates the general low level of technology in Denmark. The Danes were missing the concepts and expressions, the very language that the German academics in

Göttingen and elsewhere had developed. It speaks in favour of such an interpretation that the translator had consulted a number of artisans and craftsmen on the matter. Only to discover that they too had no words to match the German concepts and expressions.

The Institute for Metalworkers

The second point of interest was the opening of an Institute for Metalworkers in Copenhagen in 1807. This private institution was founded by a military officer, major I. Conradt, who also became manager of the institute. The purpose was to improve and heighten the skills and status of artisans in Denmark. This should be done in the workshop, where artisans could receive practical education in the long evenings of the winter. The institute opened with only 8 students. The workshop expanded during the following years. It never exerted any considerable influence on the handicrafts.[5]

Two points are remarkable about the institute. Top-ranking officials of the enlightened absolutist government supervised the institute. Among them were Ove Malling[6]. He held a chair in the board of Directors of the University and the Learned Schools until his death in 1829. The institute was incorporated into *Polyteknisk Læreanstalt* in 1832. This happened after a great schism at *Polyteknisk Læreanstalt* over the position of the workshops in the academic curriculum had occurred[7]. At the same time the workshop lost its position in the education. The utility of the workshop was not improved to a very large extent.

The example may illustrate a lack of demand for improvement of the apprenticeship. There seems to have been only little need for improved education among artisans and craftsmen in the Danish society at that time.

A Society for the Diffusion of Scientific Knowledge

In 1824 the famous physicist, professor H.C. Ørsted, formed a popular Society for the Diffusion of Scientific Knowledge. Ørsted, who had discovered and described the principles of electromagnetism in 1820, was the first director of *Polyteknisk Læreanstalt* until his death in 1851. He was a very determined and dynamic figure in the development of a polytechnical culture. Ørsted played an important part in the formation of *Polyteknisk Læreanstalt*. This institution was located partly on his premises. A lot of cooperation between this Society and *Polyteknisk Læreanstalt* went on after 1829.

The Society spread the new gospel of natural science among the common man. It undertook a range of popular scientific activities giving regular graduate level lectures on natural history in various Danish cities[8].

This society was never incorporated into *Polytechnisk Læreanstalt*. All along it was seen as a partner in a systematic division of labour. The chemist and close friend of H.C. Ørsted, professor Zeise, who was soon to hold a chair at *Polytechnisk Læreanstalt*, commented on the relationship between the Society and the coming Institution:

> "To impose the lectures (at *Polytechnisk Læreanstalt*) directly on the class of artisans and craftsmen and the dilettantes, should partly or totally be left in the hands of the Society."[9]

As a result of this formative process activities would have to be coordinated between the Society and *Polytechnisk Læreanstalt* or lumped together as would best fit the occasion.

Many of the teachers at *Polytechnisk Læreanstalt* were engaged in the activities of the Society. Some aristocrats and industrialists from the bourgeoisie displayed an interest in the activities of the society. The king patronized the Society and there was a common goodwill towards it. It cannot be ruled out though, that much of the support was caused by snobbism, rather than a profound interest in the natural sciences. Membership of the Society might lend a higher status to some people.

A Magazine for Artisans and Craftsmen

The publication of a polytechnical journal from 1826 – 1842 was a succes within certain limits. The Magazine for Artisans and Craftsmen published by the astronomer and mathematician, professor G.F. Ursin, communicated new technology. Ursin sent a proposal to the king in late fall 1827 suggesting a school for artisans and craftsmen in Denmark, modelled after the German pattern.

Ursin's proposal was the impulse that started negotiations and resulted in the opening of *Polytechnisk Læreanstalt* two years later. Ursin was appointed professor of mechanics and engineering (Teknologi) at *Polytechnisk Læreanstalt*. He resigned his post in 1832 after the scandal.

Ursin modelled his Magazine after the concept of the German magazine 'Dinglers Journal'[10]. Ursin´s Magazine held an important position as communicator of new technology from other countries to the public. Also communications between Danish inventors occurred, but at a very low rate, which indicates the low level of national invention and innovation.

The Magazine kept intimate ties with *Polytechnisk Læreanstalt* throughout the years. Most intensively while Ursin held his chair, but the link was never severed completely. On the other hand, the Magazine was connected with the Society for the Diffusion of Scientific Knowledge. Month in and month out, year after year, the magazine published a report of H.C. Ørsted's monthly lecture. Ørsted gave his lecture on the latest scientific findings to members of the Society and students at the University at the same time.

It is difficult to give a full assessment of the importance Ursin's Magazine might have had for the polytechnical culture. It held a very central position in the public. And it's a good source from which to follow the development of a Danish polytechnical culture during the first half of the century[11].

The formation of *Polytechnisk Læreanstalt* 1827-34
The political situation

The negotiations for a technological institution accelerated in November 1827, when Georg Frederik Ursin proposed the king that a school of technology be founded i Denmark. His proposal aimed at an education for the artisan and the craftsman[12].

Ursin's initiative started a series of events. The king passed the proposal to the Board of Directors of the University and the Learned Schools. The Directors submitted it to the Consistorium at the University of Copenhagen, which is the governing body of the University. Only the professors were members of the Consistorium, and some professors of science were appointed to a committee to report on the proposal[13].

Later on, in July 1828, the king submitted professor Ursin to the committee[14]. It then consisted of professors of science at the University, who were all to take up chairs at *Polytechnisk Læreanstalt* in 1829[15].

The professors attitude towards the negotiations seems to have been very pragmatic and to order, but things were also coming their way. There always exists a certain academic air in the written material they left from the process. Here the polytechnical university as a vehicle for 'polytechnical science' is advocated.

The outcome of this process of construction was almost destined to be an institution "*in the closest connection with the university*", an often recurring phrase in the documents. It was obviously the intention of the university-directors to have this outcome. And so it became a historical fact to a very high degree. The Board of Directors played a strong hand in the negotiations. On one side they were the true and humble servants of the enlightened absolutist monarchy. On the other side they had their own game of politics going on. They advocated a view on polytechnics inspired by *the École polytechnique* in Paris. There they were in unison with the professors' attitude.

As administrators of a zero-sum budget for the University of Copenhagen the Directors had almost no room to move. Still they were eager to unite the natural sciences in a faculty separate from medicine. Ursin's proposal came in handy. It could be used as a vehicle for this ambition. But the intention was never to improve the handicrafts.

In his proposal Ursin had used the term *polytechnical school*. The Board of Directors consistently talked about a *polytechnical institute*. While the committee of professors called it a *polytechnical academy*, (Læreanstalt). All this might suggest the differing views of the factions on polytechnical education.

The primary interest of the state, here embodied in the Board of Directors of the University and the Learned Schools, was a polytechnical institution for the production of candidates to be employed by the royal administration. It should be done by the lowest possible budget.

The primary interest of the professors was to assemble the sciences in a separate faculty at the university. The existing chairs, funds, and buildings could be reallocated for this purpose.

The originator of the proposal, G.F. Ursin, had preconcieved a school for artisans. He was easily persuaded into this bigger and far more ambitious project of a *Polytechnisk Læreanstalt*. Though he seems to have had his doubts about the final outcome of the process[16].

The primus motor in the whole process of formation was H.C. Ørsted, who wrote all the drafts for the reports himself. His ability to mediate between the differing interestgroups was very constructive. His own interest in the project was to create a polytechnical academy where a new experimental and applied natural science could be developed.

This is the strategic picture of the powerful groups, which created an education for polytechnical candidates in Denmark. The Directors' policy was to remove the natural sciences from the philosophical and the medical faculties and regroup them in a polytechnical institute. A faculty of science was opened in 1850, and this would give more room to *Polytechnisk Læreanstalt*. Of course this proces of restructuring the University of Copenhagen occurred while H.C. Ørsted was Rector of the University[17].

The formative years of *Polytechnisk Læreanstalt* was indeed crucial and trendsetting for the future development of polytechnics in Denmark. It was long after

H.C. Ørsted's death in 1851 that the first reform of the school came. The education of civil engineers opened i 1861 and courses in building and construction were finally included in the curriculum. In 1858 the Royal Academy of Agriculture opened, and the courses in agricultural technology were removed from the curriculum of *Polytechnisk Læreanstalt*.

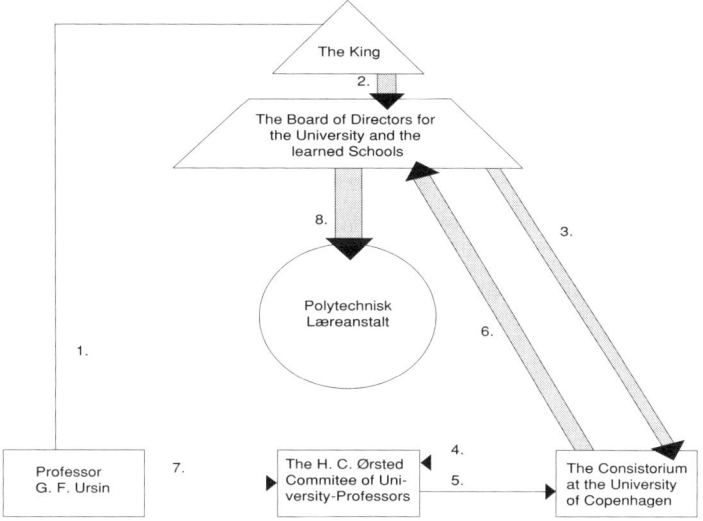

Figure 1: The chain of events that led to the formation of Polytechnisk Læreanstalt

1. The proposal to the King, 26. November 1827.
2. The King submits the proposal to the board of Directors, 22. Dec. 1827.
3. The Board of Directors submits the proposal to Consistorium at the University of Copenhagen.
4. The Consistorium appoints a Committee of University-Professors to consider the proposal and make a report to the King.
5. The Committee reports to Consistorium, 27. March 1828
6. The Consistorium sends the report to the Board of Directors, 8. April 1828.
7. Professor G. F. Ursin is summoned to the Committee of Professors. The Committee is ordered by the King to elaborate a detailed proposal for Polytechnisk Læreanstalt, 19. July 1828.
8. The King gives Polytechnisk Læreanstalt its provisional plan, 27. January 1829.

Polytechnics as applied science

How did these professors at the University of Copenhagen regard this new development in science? Now they were forced to adapt science to practical ends. The answer is, they had no regard for it at all. Everything looked almost the same to the physicist, the mathematician, the chemist, the geologist, and the botanist. They would give their usual lecture, but now also to students of applied science.

This traditional curriculum was extended by courses in mechanics (technology) and technical drawing. A workshop was also included in the institution under the auspices of the mechanic Ole Winstrup, who resigned his post in frustration in 1831[18].

The first draft of a report came from the committee of professors in February 1828, written solely by H.C. Ørsted, who played a leading part in the whole proces. The report discussed the attitude towards a polytechnical institution. And this is illuminating later events. Many of the ideas formed here were to recurr later on. The negotiations mainly concentrated on a discussion of the curriculum. The whole question of workshop-practice was postponed to later and 'informal' discussion[19].

In May 1828 the Board of Directors issued their preliminary report on the polytechnical school[20]. The report was quite extensive, and consists of three separate parts; it resumed Ursin's proposal for a polytechnical school; then the professors presented their own position; finally the directors of the university had their say. The king made this report a first blueprint for the new institution. This gives the document a particular importance in the assessment of the final result.

The report is a long argument for the specific design, actually shaping *Polytechnisk Læreanstalt*. A design that remained almost unchanged throughout the first thirty years. In other words, the formation was not only a very trendsetting process, but this document had an important function in it.

The report characterized Ursin's proposal as a school of practical teaching and education. The daily work should take place in a workshop and a chemical laboratory. Courses were given in mechanics, chemistry, technology, engineering and technical drawing. In addition to this, there should be given lectures in pure mathematics. The report here concludes by stating, that:

> "in Germany a similar type of school's is working under the name of 'Gewerbeschulen'. This kind of school is of course a useful thing for the artisan who can improve his skill, and thereby useful to mankind who will have better products. But a higher educational institution in close connection with the university would obviously be far more beneficial to the state."[21]

A new proposal for a technical university was then put forward in the second part of the report. Here the vision of the committee is outlined. The committee of professors reversed the alleged imbalance between theory and practice of the original proposal. Now the scientific courses in mathematics, physics and chemistry were seen as the backbone of polytechnical education.

The committee rejected the idea of an institution, that would:

> "simply improve the skills of the apprentice, and enable him to perform a

certain business or group of businesses by employing the forces of nature. It would be proper to communicate a higher degree of education, refinement and insight that would make the candidates useful in all enterprises employing the powers of nature. Whether it be in the state or the higher industries."[22]

The blueprint that followed was motivated by this opposite view. It was a plan for a polytechnical university modelled after the Écolè polytechnique in Paris.

Another benefit given to the Danish state from this construction would be the teachers forming a technical commision, which could be consulted by the government in questions of technology. This idea was actually one of Ørsted's original inventions. The teachers at *Polytechnisk Læreanstalt* were consulted by the state administration, and had to function as board of technical advisers. This was probably the most practical effect that *Polytechnisk Læreanstalt* had on industrial enterprises. And it served a strictly bureaucratic function. In the years to come it was the basis for decisionmaking in cases where a person applied for the grant of a privilege to embark on a specific new industrial production. The project would then be evaluated by the technical board.

Admission to the polytechnical education

The question of admission to *Polytechnisk Læreanstalt* was crucial to the negotiations. To be admitted to *Polytechnisk Læreanstalt* one should be academically prepared. First you had to pass the philosophical-philological exams at the university. Alternatively, you had to be an officer with the proper exams from the army. The participation of officers in this new polytechnical education never became very impressive, as the army had its own system of education. In 1830 the Royal Military Academy of engineering opened. The majority of the officers were educated here[23].

Of course some overlap between military and civil education had to take place, due to the fact that until 1857 building and construction could only be studied at the military academy[24].

Polytechnisk Læreanstalt was designed to produce two types of candidates. They were named candidates of applied science and candidates of mechanics. In addition to this students from other faculties followed lectures at *Polytechnisk Læreanstalt*. Also people who were interested in science, but indifferent as to pass any exams, were admitted for a fee.

The candidates of applied science studied the traditional curriculum of the mathematical, chemical and physical sciences. These courses were supplemented by courses in technology, mechanics, engineering and drawing. This was inspired by Ursin's original proposal. The candidates in mechanics had a larger portion of mathematics, technology, mechanics, engineering and drawing. The mechanical students still spent a great deal of time studying the traditional curriculum of natural sciences.

A reform work started very slowly after Ørsted's death in 1851. A broader design for the education of civil engineers was gradually enacted from 1859. The reform was made to meet the new needs of the public services. Here a growing number of civil engineers were needed to construct, build and run the growing number of public works.[25]

The workshops

Concerning the establishment of a workshop practice at *Polytechnisk Læreanstalt*, the visions of the committee became more blurred. The professors suggested without great conviction, that the polytechnical student would benefit from learning workshopskills. Of greatest value to the student was the workshop because:

> "it enabled young people of fine upbringing and cultivation to have the opportunity to learn the different skills and trades, without having to undergo the ordinary apprenticeship so inconvienient for those young men in question."[26]

Of much greater importance was the intimate connection with the university. This point is stressed again and again in the report. Refering to the University of Prague, the report suggests that this connnection will benefit the university scientifically in new and fruitful ways. Moreover, there were very considerable savings in expenditure to be had from the plan. The intention was to reallocate existing funds and spend them on this new purpose[27].

The Board of Directors had the final say in the report. They accepted the new proposal for a polytechnical university. Perhaps this is not very surprising. The Directors concentrated on many problems having to be solved, a serious blow to the original proposal had already been delivered. This was designated as "*narrow*", or perhaps "*narrowminded*" is the correct translation here. The professors also claimed for their part, that "*it was common knowledge, that these schools for artisans and craftsmen were not beneficial to the state at all*".

Actually, these harsh remarks from the professors would provoke the intervention by one of the Directors, Ove Rothe. He gave a serious reminder to the Consistorium in the report. These schools certainly did have a very positive influence on the artisan and his trade. Compared with the new proposal for a polytechnic university, the original proposal was of less interest, because this new plan would be very beneficial to the state itself, simply because it was based on science. By this conclusion the strategy for the development of a polytechnical institution in Denmark was literally chosen. It was developed starting from the top rather than from the bottom as the original proposal had intended.

Maybe the king cared just a little bit more about the improvement of apprenticeship than the professors at the university did. The small indiscretion from Director Rothe may show us the breech in a united front against the handicrafts. The intervention was probably wise also, because there had to be a mediation between different views and interests in this matter. If some practical results were to evolve out of all this thinking, all the arguments must be considered in the proces, while the final decision rested in the highest office by the king.

The Royal administration was very pleased with the report as it came out. In a resolution of 19th July 1828 the king appointed the same committee of professors to make a new report. The originator of the proposal, G.F. Ursin, was ordered to join this committee. The professors were now to be dealing with the practical problems in the blueprint of *Polytechnisk Læreanstalt*. Their main concern in planning this academic institution, was its housing, its financing, its organization and scientific connection with the university[28].

Danish polytechnics between handicraft and science

The committee of professors were in session for the next half year making the final plans for the institution. A lot of paper was going back and forth between the professors and the directors of the University. The directors would cut the proposed expenses to a minimum. In the middle of January 1829 the final report was sent over from the Board of Directors to the king. By a rescript of the 27th January 1829 the king gave *Polytechnisk Læreanstalt* its statute of foundation[29].

This document plainly stated the basic purpose of the provisional plan for *Polytechnisk Læreanstalt*:

> "The purpose of *Polytechnisk Læreanstalt* is to communicate to the young people, who have the necessary good foundation, and such an insight into mathematics and experimental natural science and such a skill in the use of this insight, that they hereby would become extremely useful in certain sectors of the state administration, also they may be useful in running industrial plants."[30]

In the statute the priorities of the state were openly declared. When the new institution opened in the prescence of the king, Ørsted flatly stated the order of priorities as the most natural thing in the world. He also discussed the benefits for industry:

> "It is not unknown to me that many of the foremost and most enlightened men in this country suffer from the false imagination that scientific knowledge is detrimental, even damaging to the tradesman."[31]

But this was not the case of the natural sciences in Ørsted's opinion. The experimental sciences lead to action. Certainly, to study the science was not a job for everybody to do, but the benefits were for everyone. From the study of science stems the common good:

> "When the scientist discovers the many things, which the theory has been unable as yet to explain, he is forced to new experiments that will now correct and now enlarge his insights; while he, on the other side, enjoys a new and beneficial pleasure from his science, as he can see, how it embellishes the bourgeois life."[32]

The polytechnical institution was to fulfil its designated purpose, as the cameralist administration had preconcieved it. During its first thirty years of existence *Polytechnisk Læreanstalt* produced 131 polytechnical candidates. Out of these 70% went to work in the public sector. Be it the royal administration, teaching at graduate and post-graduate levels, or the construction, building and management of railroads and public works. Another 29% of the candidates went into industry or farming. They were often the sons of a factory-owner. Or it could be candidates working as estate-managers who then married into bigger estate. The last 1% died early, were idle or went abroad. According to type of education, the figures are as follows:

Distribution of employment for polytechnical candidates 1831-1861

	Public service	Private industry
Applied science candidates	39%	24%
Mechanical candidates	31%	5%
Total	70%	29%

(Source: J.J. Voigt, *Statistiske Oplysninger angaaende den polytechnisk Læreanstalts Kandidater*. Kjøbenhavn, 1890.)

The figures indicate that a Danish industrial technology did not go anywhere at all until long after 1840. Industrialisation never went very far at all in this period from 1831-1861 which is best described as the take-off phase of an early industrial development[33]. The demand for polytechnical candidates in private industry was limited by the small scale development of industry. The distribution of the candidates on employment speaks for itself. A strong cameralist tradition in the state prevailed. Very few new industries emerged during the first half of the 19th century and they had little requirements for a science based technology like the one developing at *Polytechnisk Læreanstalt*.

The professors at *Polytechnisk Læreanstalt* did not suffer from any illusion, that scientific knowledge was damaging to the trades, the artisans and the craftsmen. However, they were running a polytechnical institution strictly tuned to another purpose, but this didn't keep them from caring about industrial obligations in the leisure time.

The Director of *Polytechnisk Læreanstalt* from 1851-1865, J.G. Forchammer, wrote and published the invitation to form the Industrial Association in Copenhagen in May 1838[34]. This association was to become a forum for both polytechnical candidates and industrialists. Hummel also shared the burden with G.F. Ursin of publishing and editing the *New Magazine for Artisans and Craftsmen* from 1836-1839.

A quite different type of technological society was formed in 1848 by would-be professor of mathematics and official historiographer of *Polytechnisk Læreanstalt*, Adolph Steen. After the fall of absolutism, Steen took the initiative to form the Polytechnical Society, which held 12 meetings until it finally folded in 1851[35].

The primary purpose of this professional organization of polytechnical candidates, was to make a claim on the justified use of scientific and mathematical education in the State. The Society referred directly to the political conditions under the new democratic administration, that took power after the fall of enlightened absolutism. This bourgeois revolution removed the former garantees and privileges of the public servants.

It was the first attempt of the polytechnical candidates to create a professional organization for 'civil-engineers' in Denmark. But it was stillborn. When this Society did not amount to very much, it might stem from the fact that the democratic state administration also needed a growing number of public servants with an education in polytechnics. Although the new democratic constitution would not grant privileges to any specific class or group or profession in society, the polytechnical candidates soon gained a prominent position in the public services, without any help from the Society. A professional association for engineers (DIF) was established in 1892[36].

Conclusion

Polytechnisk Læreanstalt was developed as a vehicle to promote and apply science to the administration of the state and the public services in Denmark. The necessities of industry came second or third in this priority. This does not mean that nothing concerning industrial technology happened at *Polytechnisk Læreanstalt*. It just meant that the primary objective of the institution was to produce administrators, not to educate or even promote new scientific engineers for the nascent business and industry.

Still, new technology was promoted by this institution. And new technology was even developed inside its workshop. Although in a very small scale.

In conclusion I should like to present an exceptional case of technological development. It was so unusual that the best way to characterize it would be as a Weberian idealtype. Had this phenomenon not been an actual case.

This historical case connects the disparate sectors, that were discussed in the beginning of this paper as early sources of a culture of technology. This particular piece of new technology was communicated by the Magazine for Artisans and Craftsmen. The inventor, mechanic and industrialist, Ole Winstrup, was at that time still in charge of the workshop at *Polytechnisk Læreanstalt*. In an article Winstrup described his piece of invention in great detail. And it was illustrated by a technical drawing[37].

The idea of this invention was given to Winstrup by Ursin who had suggested the construction of a pressure pump to be used at *Polytechnisk Læreanstalt* for multiple purposes. It was to serve as a fire pump at the institution. Normally, it was used every day to pump water into a tank on the 2nd floor of the building. The pump was operated by handpower and enabled the chemical laboratories at *Polytechnisk Læreanstalt* to have running water at their disposal. Needless to say, the daily duty of pumping water to the laboratories was used as a disciplinary punishment of the students.

Admittedly, this is a very special case of new technology. Still, it might not be very far from an ideal position of Danish polytechnics, as envisioned by some of the founding fathers. It was motivated by the visionary struggle to create a polytechnical science literally from scratch and give it a prominent position in society in between the handicrafts and the experimental natural sciences.

The founding fathers of an academic Danish culture of technology seemed to be perfectly capable of creating such a practical project. At least, when they were not too busy being engaged in the theoretical education of polytechnical candidates for the state and the public services.

Notes

1. I am most grateful for the inspiration, though not in accordance with the conclusions put forward by Dan Ch. Christensen in his paper: "Naturvidenskabelig og teknologisk udvikling i socioøkonomisk sammenhæng. En analyse af H. C. Ørsteds virke", in *Humanistisk Årbog* bd. II. RUC, 1986. As it seems to me, that Dan Christensens critical perspective on the formation of Polyteknisk Læreanstalt was more or less generated around 1890, without taking the situation in 1829 into serious historical consideration.
2. Johann Beckmann. *Anviisning til Technologie, eller til kundksab om Haandværker, Fabriker og Manufacturer, fornemmelig dem, som staae i nærmest Forbindelse med Landhuusholdningen, Politie- og Kameralvidenskaben. Tilligemed Bidrag til Kunsthistorien. Oversat efter fjerde og forøgede oplag.* København, 1798.
3. Cf. Axel Nielsen. *Den tyske Kameralvidenskabs Opstaaen i det 17. Aarhundrede.* Det Kgl. Vidensk. Selsk. Skrifter. København, 1911.
4. Cf. Axel Nielsen. *Det statsvidenskabelige Studium i Danmark før 1848.* København, 1948.
5. H. P. Selmer: Academiske Tidender, eller Samling af Efterretninger vedkommende Kjøbenhavns Universitet, Sorø Akademi, og de lærde Skoler, samt hermed forbundne Anstalter. 2den Aargang, 1834. pp.476-481.
6. Ove Malling, 1748-1829. Top-government official and royal historian. 1783-1802 president of the Royal Board of Agriculture. From 1805 memeber of the new Board of Directors of the University and the Learned Schools, after 1810 1.member of the Board. Here he took a great interest in the reconstruction of the cameralist education at the Sorø Academy, which reopened in 1821. From 1823 head of the Royal Library. From 1813-18 Director of the Royal State Bank. On January 9, 1924 he was appointed *Prime-Minister*, and together with A. S. Ørsted (younger brother of H .C . Ørsted) chaired as the only members of the *Prime-Council* of bourgeois ascent. Malling and A. S. Ørsted both had refused knighthood. Ove Malling is generally biographed as a very sympathetic and openminded person, with an enormous working ability. It is said that he was able to communicate with all sorts of people from every class in society. C. F. Bricka, (ed.) Da. *Biografisk Lexicon. XI.* København, 1897.
7. J. T. Lundbye. *Den polytekniske Læreanstaltt 1829-1929.* København, 1929. pp.70-78. The general student attitude appears to be very negative towards the practical training and applied science courses. Only the chemical laboratories drew much attention from the students. Professor G. F. Ursin was made responsible for the general failure of the courses in drawing, technology and mechanics. Out of 6 students only one passed the exams in December 1831. As a result of this Ursin was forced to resign in February 1832. This didn't improve the position of the applied sciences and the workshop *Polytechnisk Læreanstalt*. The combination of experimental science and workshop practice continued to create problems for the academic curriculum.
8. M. C, Harding. *Selskabet for Naturlærens Udbredelse. H. C. Ørsteds Virksomhed i Selskabet og dettes Historie gennem hundrede Aar.* København 1924.
9. PLA./1828: No.2 Zeise til Ørsted. 1. Marts 1828./RA
10. *Polytechnisches Journal.* Herausgegeben von Johan Gottfried Dingler. Erste Band. Mit einem Vorwort von Wilhelm Treue. Reprint. New York, 1969. There were three sources of inspiration that went into Dingler's Journal: a) Beckmann's Technologie; b) French natural scientific literature; c) British experimental polytechnics.
11. Cf. Michael F. Wagner. *Registrant over Ursin's Magazin for Kunstnere og Haandværkere.* Upubl. Historisk Institut, Aarhus Universitet, maj 1991.

12. Dir. f. Univ. og de lærde Skoler; Forestillinger 1829, No. 1155. 1. D. 215 a. Til Kongen! Letter from G. F. Ursin. 26. Nov. 1827/RA; Polytechnisk Skole. 26. Nov. 1827. Ibid., 1. D. 215 b./RA
13. H. C. Ørsted gave a first description of the procedure in the first committes report from March 27, 1828. It is consistently repeated in later reports; PLA\ 1828: No. 3. 27. Marts 1828./RA; Dir. f. Univ. og de lærde Skoler, Forestillinger 1829, No. 1155. 1. D. 374. 8. April 1828. Fra Consistorium til Directionen for Universitetet./RA.
14. Dir. f. Univ. og de lærde Skoler, Forestillinger 1829, No. 1155\ 1. D. 477. Kgl. Resolution af 19,Juli 1828, om Oprettelsen af en polytechnisk Læreanstalt./RA.
15. Except professor Thune, who died 11. Arpil 1829. cf. C. F. Bricka (ed.) *Biografisk Lexicon, Bd. XVII*. København, 1903; Thune was replaced in the comittee by J. G, Forchhammer in August. Dir. f. Univ. og de lærde Skoler, Forestillinger 1829 No. 1155\ 1. D. 502. 19. August 1828./RA.
16. In an undated letter probably from late 1828 Ursin expresses his severe doubts as to whether the project will ever become reality. Ørsted in response had claimed, "that they should congratulate each other instead of wooyring, because the outcome was certain". PLA\1828. No. 8./RA.
17. Mogens Phil, ed.: Københavns Universitet 1479-1979, Bd. XII *Det matematisk-naturvidenskabelige Fakultet 1. del*. København, 1983. pp.33-74.
18. P. L. A\1831: No. 9a, 9b; 28. Januar 1831, and No. 11a, 11b, 11c; 1. Marts 1831./RA.
19. P. L. A.\1828: No. 1. 21. Februar 1828. Betænkning til D*Herrer Professor Thune, Zeise og v. Schmidten fra H. C. Ørsted. 1828\RA; Compare with the provisional Plan for polytechnisk Læreanstalt, and provisional Regulative for polytechnisk Læreanstalt: P. L. A. Reprinted in A. Steen. *Den polytechniske Læreanstalts første halvhundrede Aar 1829-1879*. København, 1879. pp.106-118.
20. Dir. f. Univ. og de lærde Skoler, Forestillinger 1829 No. 1155. 1. D. 374. 22. Mai 1828. angaaende en polytecknisk Skole omtalt. Allerunderdanigst Betænkning./RA.
21. ibid.
22. Dir. f. Univ. og de lærde Skoler, Forestllinger 1829 No. 1155. 1. D. 374. 22. Mai 1828 angaaende en polytecknisk Skole omtalt. Allerunderdanigst Betænkning./RA.
23. Cf. V. E. Tychsen. *Fortifications-Etaterne og Ingenieurkorpset 1648-1893*. København, 1893.
24. Cf. Adolph Steen op. cit.
25. Cf. J. T. Lundbye. Den polytekniske Læreanstalt. København, 1929. pp. 98-136, esp. p. 121 ff.
26. Dir. f. Univ. og de lærde Skoler, Forestillinger 1829 No. 1155. 1. D. 374/RA. This is a coomon phrase in the responsa and reports to appear after the question is considered informally some time in March 1828. It certainly appears in the Report of May 22. 1828.
27. This idea already appears in Ørsted's first report of February 21st 1828, op. cit.
28. Dir. f. Univ. og de lærde Skoler, Forestillinger 1829 No. 1155. 1. D. Kgl. REsolution af 19 juli 1828./RA.
29. Reskript (til Directionen for Universitetet of de lærde Skoler) ang. Oprettelsen af en polytechnisk Læreanstalt og Approbation paa Planen til Samme. In Algreen-Ussing, ed., Kongelige Reskripter og Resolutioner. VII. Deel, 12te Bind. 2det Hefte. Kjøbenhavn 1835. pp.45-55.
30. Op. cit. p.48-49.
31. H. C. Ørsted. Om den dannende Virkning Natervidenskabens Anvendelse maa udøve. In Aanden i Naturen. 2. Deel, Kjøbenhavn, 1850. pp. 202-212, (my transl. mw). Engl. transl from the german transl. The Spirit in Nature. London, 1966.
32. ibid.
33. Ole Hyldtoft. *Københavns Industrialisering 1840-1914*. Herning, 1984.
34. Nyt Magazin for Kunstnere og Haandværkere. Vol. 2. 1838, No. 110, 31. Mai, p.398-400; No. 111, 7 juni, p. 401-404.

35. Polytechnisk Samfunds Forhandlinger. Protokol 18/12 1848- 10/11 1851. PLA./RA.
36. Cf. Morsing. *De ansete Mænds Fagforening. Dansk Ingeniørforening 1892-1992.* København, 1992; The polytechnical candidates and the engineers didn't form a professional organisation until 1892, when DIF (Danish Association of Engineers) was founded. This formation was due to the growing demand for engineers in private industries; *Ingeniøren* Vol. 1. No. !, Juli, 1892; DIF here stated their three purposes for taking up activities as: a) promoting and improving the polytechnical education; b) claiming the position of engineers in the public services; c) fighting for better wages and work-conditions in the private sector.
37. O. J. Winstrup. *Trykpumpe, udført af O. J. Winstrup. (Hermed Tegningen Tab. XII.)* Magazin for Kunstnere og Haandværkere, 2. rk. No. 1 p.446-449. København, 1830.

The Danish Engineer in Transition – The Reformation of Danish Engineering Education c. 1890-1933

Henrik Harnow

Introduction

This paper deals with the successive attempts to modernize the somewhat insufficient engineering education which was offered by the only technical university in Denmark during the 19th century, Polyteknisk Læreanstalt (PL).

I will mainly focus on the two important reforms of 1894 and 1933, concentrating on the problems of theory versus practice, taking into account changes in curricula, physical facilities, teaching and international orientation. Lastly I will take a look at the job situation in engineering in 1870, 1890 and 1930 trying to reach conclusions about the practice versus theory problem in relation to the suitability of Danish engineers to the needs of their times.

The problem of practice versus theory is, I believe, a tricky one: practical subjects such as hydraulics are also treated theoretically. In this connection I understand by theory the teaching of the sciences. Though the practice v.s. theory problem is important, I also believe that other important parametres must be considered in order to reach meaningful conclusions in an aspect of engineering history that often takes on an ethereal character.

Prelude

The first formal high-level engineering education in Denmark originates from the founding of the first technical university, the PL, in Copenhagen in 1829. This institution was founded and heavily influenced by the Danish physicist Hans Chr. Oersted, of magnetic induction fame.[1]

The PL gained a distinctly Danish flavour from its connection with Oersted and the University of Copenhagen. Though a technical high school with the aim of educating young men suited for industry, during its early phases the PL was gravely lacking a practical side. It is no overstatement to conclude that the Danish technical high school between the founding year of 1829 and the introduction of the first genuine engineering curriculum in 1857, primarily educated technical teaching staff.[2] The PL was a very theoretical school, a reputation it kept long after its initial phases.

It must be remembered, though, that the industrialization of Denmark and the rising need for engineers, was of a relatively late origin seen in an international context. Only from the middle of the 19th century did industry gain ground, real industrialization taking off during the 1870's and finally reaching full speed around the 1890's. Although the PL is an easy target for criticism during its early years, when the theoretical abilities of an engineer were of minute importance contrasted with statical intuition and a practical background, it must not be forgotten that Danish industry was not really in a position to gain from engineering abilities until the 1870's.

After establishing a course in civil engineering i 1857, the curricula of the three courses, chemical, mechanical and civil engineering, were relatively unaltered until the reform of 1894. The civil engineering course was a big step forward and soon became the most popular one. One deficiency is obvious, though, in that civil engineering in all its aspects were covered by only one man, the harbour constructor of Copenhagen, Mr. L. F. Holmberg, between 1857 and 1894.

In 1884 the basic pattern was established with a first theoretical part taught by the university staff and a second "practical" part. The practical aspects were in fact quite limited, except for the chemists. The mechanics and the civils went on a few excursions, watched experiments now and then but mainly concentrated on drawing, spending as many as 12 hours weekly during some semesters.[3]

The 1870's was still a time, when mechanical laboratories were in their infancy on an international level. Engineering had still not taken up the research role, which later proved so extremely demanding of space. Everybody, not least the school itself, was aware of the acute limitations during this period, the physical facilities, some old converted professors' dwellings, which for years had been a real obstacle.

The teaching staff no doubt was very competent, and the quality of the teaching of the sciences has never been critizised. Criticism has only been put forward on the internal balance between the sciences and real practice.

Finally, after a long struggle caused by the political conflict during the years of the socalled Withering Policy, plans for a new institution were finally approved. After a short but hectic period of building activity, a new school complex was taken into use in 1890. It will suffice here to say that it was mainly in terms of more space and of larger and better lit rooms for drawing practice that the new buildings were an improvement. Their construction mirrored the fact that by 1890 the PL still viewed engineering education as basically meant to take place in a traditional classroom.[4]

Before we move on to discussing the reform of 1894, we must take a look at the use made of Danish engineers during these years. Statistics gathered in the 1870's indicate that the engineers from PL were doing quite well by then, no longer being confined to teaching. We find quite a few municipal engineers and also engineers involved in industry. It seems that after a short period of apprenticelike connection with a firm, many engineers were able to take over more demanding jobs.[5]

Before celebrating these facts, we must note that all was not well. Most really demanding engineering projects of the time were being carried out by foreigners, mainly English, but also German and French engineering firms.[6] The Danish railways, by far the most important engineering project of the 19th century, were being built by the firms of Fox, Henderson & Co. and Peto, Brassey and Betts. Also the gas- and waterworks of the 1850-60's typically involved English firms. As late as 1879 the French firm Compagnie de Five built the railway bridge across Limfjorden in Jutland.

Times were changing. By the 1880's the Danish entrepeneur Niels Andersen built 83 km of railway in Sweden, and foreign contractors lost importance. It seems that Danish engineers had finally learned their trade by then. Also noticeable changes in engineering were taking place, the engineering sciences taking on importance. This proved to be a real advantage to the theoretically founded Danish engineers from the PL.

The reform of 1894

During the 1880's dissatisfaction with the engineering education at the PL was mounting. Still, it was not until the foundation of the first genuine engineering organization in Denmark in 1892, the Association of Danish Engineers (DIF), that the debate really took off.

The new organization, which was aimed primarily at candidates from the PL, and from the start numbered several distinguished engineers among its members, immediately took a stand on the issue of education. In fact one of the chief aims of the new organization was to "...raise the importance of scientific technical education...".[7]

The critics outwardly acted as single individuals, but the DIF at the same time functioned as a coordinating body in organizing meetings on the subject. It also lobbied within the Ministry of Education by submitting proposals for a revision of the PL and by offering to take part in a commission itself.[8]

The critique, to which I shall return shortly, was no doubt competent and put forward by some of the most eminent engineers of their time, but it also served as a vehicle for the new organization by which to gain influence. In the end this critique also served the teaching staff of the PL well, not only in that extra staff were employed, but also by raising salaries. Conclusively the new organization proved both successful and influential right from the start.

The critique can be summarized in 4 main points: 1) the PL needed a general revision, 2) more teachers were necessary, 3) better wages for teaching staff were necessary and 4) a committee had to be created to take responsibility for a revision of the curriculum. This overwhelming critique was backed up by individual contributors to the new journal the *Engineer*, with examples drawn from "the real life of the engineer" – industry.[9]

Most critics agreed that it was not feasible that one man – Holmberg – could master the whole spectrum of civil engineering all by himself. Civil engineering had become the most popular branch of engineering at the PL, and by 1890 the intake of students was steadily rising. In Germany, it was pointed out, several professors and assistants covered a similar area.

All praised the quality of the scientific teaching, the level of which was considered high – too high in the opinion of some. A major point was the need to include more practical aspects, especially in mechanical engineering, which suffered from outdated books and unsatisfactory facilities. It was pointed out that in Germany both Hannover and Munich possessed far better facilities and a larger teaching staff.

A former candidate from the PL, by then a military engineer, described how incompetent he had felt when leaving the institution, and a leading municipal engineer still thought the candidates best suited for teaching; even their drawing abilities he considered mediocre. However, it was an unfair criticism that by 1892 the PL still primarily produced teachers.

The critics elaborated on these aspects with one major aim in view: they wanted the inclusion of laboratory work or somehow to incorporate practical experience.

The debate summarized here proved quite influential, but the DIF commission, which presented these proposals to the ministry, probably made one grave mistake. They actually meant *laboratory work*, but they called it *work-shop practice* instead. If the proposal for a real mechanical laboratory had been carried out in 1894, it would have put Denmark on a par with the general level of the international development at the time. I shall return to the implications of this mistake later on.

During these years, when no education in electrical engineering existed in Denmark, and the famous technical high-schools in Germany, Switzerland and the USA exerted a strong influence, several Danish candidates studied in other countries or just visited these institutions. It was mainly to gain in practical abilities, but quite a few Danish engineers found jobs in for example the USA and some stayed for years.[10]

When it came to the question of theory, the PL and its candidates were in no doubt about their own qualities. Mr. Lütken, a teacher in civil engineering after the reform of 1894, wrote regarding his experience of America that Danish engineers normally expected "...the average American engineer to be ignorant of even the simplest theoretical subjects."[11]

The new plan for educating engineers at the PL was undoubtedly a major improvement.[12] The teachers' council produced a detailed proposal to the ministry, very much along the lines prescribed by the DIF. The council agreed that a practical aspect had to be incorporated into the courses. It was now acknowledged that the PL ought to maintain a high theoretical level, but the practical abilities of the would-be engineers were also to have a high priority.

For each branch of engineering new improvements took place. Both the socalled factory engineers (chemical engineers) and the civils faced a slightly altered curriculum, the latter now including for example materials testing and being covered by two teachers. The most substantial revision was of the mechanics' curriculum, which now included static graphics, knowledge of machinery, the science of iron shipbuilding, heating and ventilation as well as knowledge of electrical instruments and methods of measurement. For all three branches the practical element gained a little ground.

A very drastic change was the inclusion of one year of compulsory work-shop practice for mechanical engineers between the first theoretical part and the practical second part of their education. The DIF proposal had been taken at face value: the mechanics now had to spend one year in a machine-shop in the middle of their education. This meant that a degree in mechanical engineering took 5 ½ years, the others only 4 ½ years.

Eight teachers plus five university attached professors made up the new staff; this was only one less than the number requested.

A new reform could not stand alone, and the new general policy of including more practice fell on hard ground the first few years. There was no mechanical laboratory, no construction laboratory or facilities to offer a more comprehensive teaching in the electrical aspects of engineering.

In principle, at least, the 1894 reform was the beginning of the period of the modern Danish engineer and must be acknowledged as such. The physical facilities, although an improvement on the earlier situation, still did not differ from the principles of nearly seventy years earlier: the buildings were primarily meant for classroom lectures.

Between reforms

The time of the reform marks a turning point in many ways. New directions in engineering were emerging and engineering science was growing into maturity. In a more local context, the intake of students also marked a turning point at the PL, where only 480 candidates had been produced prior to 1890. Between 1890 and 1930 some 3500 passed through.[13]

The main themes of this period are 1) The 1890-school complex immediately proved too small and oldfashioned, 2) new directions in engineering demanded new courses, 3) the emergence of scientific engineering and research labs and 4) the appearance of a medium-level engineering education aimed at artisans.

The boom in student intake explains why the new school complex immediately proved too small. Soon some provisional drawing rooms had to be taken into use as well as the building of temporary barracks took place. The yearly intake of students, ranging from 15 in 1890 and rising gradually to reach 178 by 1930, was seen as sufficient for a country the size of Denmark, and with the coming of the economic crisis and the unemployment of the interwar years, no further growth was desired nor expected. A new school complex would not be constructed to cater for further students.

The developments in electrical engineering were evident, and in 1903 a new course in electrical engineering was started.[14] A temporary electrical laboratory was installed in some converted basement rooms, but plans for more wide-ranging additions were going on. Professor Hannover of the PL was deeply involved in planning the new laboratories of 1906, which included a mechanical laboratory with a boiler room as well as a modern electrical laboratory. The best German, Swiss and Danish electrical equipment was installed, delivered from e.g. Brown, Boveri & Cie., Siemens-Schuckert and Varta. Of the Danish firms Burmeister & Wain as well as Thomas B. Thrige were known ouside Denmark.

Professor Hannover's international horizon was impressively wide. He corresponded with or visited Germany, Switzerland, USA and England. Only in mechanical engineering did the PL show any interest in British matters, Hannover also showing some interest in the educational system. The answer from an English professor gives an impression of why this interest quickly vanished. The professor wrote

Dear Professor Hannover.
I am not surprised, that you do not understand our
English system of education & neither does any one else,
for there is no system.[15]

Having finally reached international standard, the PL whose director was Mr. G. A. Hagemann, known as an outstanding Danish industrialist, soon tried to solve other deficiencies, which kept appearing or reappearing during these turbulent years.

The tendency to specialize and the need for research facilities broke through with full force during the first two decades of this century. We can follow the developments by citing the following list of events at the PL: In 1906-7 the laboratory for biochemistry was opened, in 1909 low current engineering was made compulsory in electrical engineering and P. O. Pedersen became the world's first lecturer to take a chair in this field. Provisional laboratories for telegraphy, telephony and a materials laboratory was installed in rented premises in 1908. That both a stable and a former military depot were taken into use as a structural laboratory and a technical laboratory respectively, tells us what conditions prevailed during these years. It is obvious that the PL was not geared towards continuous change yet, and this halted modernization somewhat.

Nevertheless, it is impressive that Danish engineers were able to establish themselves internationally during these decades in areas of structural engineering, shipbuilding and electrical engineering. To a large extent this must be credited the very competent teaching staff at the PL, including names as P. O. Pedersen in electrical engineering, H. I. Hannover in mechanical engineering and Professor Ostenfeld in structural engineering. Also, the fact that the candidates were well qualified theoretically, proved to be a real advantage during these years, when other technical institutions altered their curricula from being primarily practical towards a more scientific character.

It must also be noted that in 1916 a technical doctor's degree, the dr. tecn., was instituted, placing engineering in a position equal to the university.

In 1929, after years of wasteful struggle over a plot of land selected for a new institution, the PL could finally lay down the foundation stone on its centenary.[16] The new school and research complex was built over the next 25 years, and though it is impossible to go into any detail here, the large sums needed for this were mainly used on a wide variety of research facilities. Being the only university level technical education in Denmark, the PL had to cover all aspects of engineering, another explanation of the problems of steering through turbulent times.

The last important breakthrough which I must mention, was the emergence of a medium-level engineering education, which by catering mainly for artisans and by aiming directly at industry, took some air out of the critique of the PL. The socalled *teknika* were started in the then third largest Danish industrial town, Odense, in 1905, but spread to several other towns between 1915 and 1933, by then covering both structural, mechanical, electrical engineering and shipbuilding.[17] They followed very closely the pattern of the German medium-level technical schools, also named teknika. The first teknikum in Odense chose the well-known teknikum in Mittweida, near Chemnitz in Saxony, as its model. Not until 1939 were the teknika-educated engineers accepted as *de-facto* engineers by the DIF, a conflict I will not treat in any detail here. It must be noted, though, that the teknika-engineers were not accepted as members of the DIF, but instead formed their own organisation, the I-S (The Union of Engineers).

The birth of modern engineer: the reform of 1933

Once again a new school complex was being built prior to the changes in the engineering education itself. This time it was necessary to have at least some of the new facilities available before revision took place. The reform could only be carried out to full extent in a combination with new facilities, which included several research labs.

Like in the early nineties the reform had been a long time coming. Already in 1914 a committee at PL had started planning the revision, but in 1926 it found its own proposals so wide-ranging that it refused to take on responsibility alone. Representatives of the DIF and the Council of Industry were included in the council, and the final plan was presented in 1932.[18]

This clearly shows that while industry had been in a struggle some 40 years earlier to influence decisions, the times had indeed changed. The accept of industry and the DIF was now considered essential, and while engineering theory had gained further importance, advice from the outside – practical – world was now being taken into account.

I have no time here to go into details concerning the single courses, but neither is this necessary to point out the main changes: these were mainly structural in that the new governing principle of engineering education was finally acknowledged: flexibility and everlasting change.[19]

Engineers on the job in 1870, 1890 and 1933

Let us now, as a small appendix to the history of engineering education, take a look at some statistics compiled for the period.[20]

Years	State etc.	Private enterprise	Teaching	Working abroad
1870-79	58%	42%	12 %	8%
1890-99	42%	58%	4 %	14%
1920-29	25%	75%	4 %	21%

It is obvious that from the 1870's there is no evidence of Danish engineers being best suited for teaching. Industry gradually accepted an initial period of apprentice-like employment, after which the engineers took over their "real" jobs. After the turn of the century the polytechnics became an absolute necessity in electrical engineering, construction and chemicals to name but a few.

During the first decades of the 20th century Danish engineers established or made up an important part of firms which became known worldwide. In structural engineering it was firms like F. L. Schmidt, Christiani & Nielsen, Monberg & Thorsen and Kampsax, which left their stamp all over the world, building harbours and docks, concrete bridges and constructing major railway lines like the more than 1000 km of the Transiranian Railway built by Jørgen Saxild of Kampsax in the 1930's.[21]

Conclusion

We have now traced the developments in curricula, teaching staff and physical facilities and contrasted this with the jobs offered Danish engineers.

Until the turn of the century, the only technical university in Denmark did not speak the same language as the engineers employed in industry, some of them being candidates from the PL themselves. It was the well known problem of theory versus practice, which occupied most of the critics.

Generally the reforms and successive revisions came too late, changes always halting a few years behind countries like Germany and the USA, which were the main sources of inspiration.

An understanding of the need for facilities that offered more than just simple classrooms was also lacking. By contrast, the teaching staff was well qualified, and the teaching of the sciences was of especially high quality. Basically, though, it was by sheer luck that the theoretically well founded Danish engineers suddenly found themselves in the forefront in areas as for example reinforced concrete construction during the first decades of the 20th century. PL had been a very theoretical high-school all along, and it was primarily external developments in scientific engineering, which were responsible for the favourable change which took place.

On this background it now seems a little ironic that the practical Danish engineers fought a battle for more practice at the same time as the candidates from the PL finally gained from their theoretical teaching and when most other technical high-schools were moving in this direction.

Notes

1. For a more detailed account of this period see Michael Wagner's paper included in this volume. Lately two volumes dealing with Ørsted's life in general and the important role that he played during the first half of the nineteenth century have been published: F. Billeskov Jansen et al. (ed.), *Hans Christian Ørsted*, Copenhagen, 1987; and Dan Ch. Christensen, *Oerstedt – A Romantic in Science and Technology*, in C. Landström (ed.), *Intellectuals reading Technology – proceedings from the Nordic Synposium "Technology, Ideology and Culture"*, 1991
2. See P.O. Pedersen, *Den Danske Civilingeniøruddannelse*, særtryk af Ingeniøren, 1937, nr. 30, 1937, p. 2. Between 1832 and 1869 at least 26% of all candidates took up teaching.
3. A. Steen, *Den Polytekniske Læreanstalts første halvhundrede Aar*, Copenhagen, 1879, p. 37-73 (covering the years 1849-79) and and J.T. Lundbye, *Den Polytekniske Læreanstalt gennem Hundrede Aar, 1829-1929*, Copenhagen, 1929, pp. 98-190.
4. This statement was put forward by the director of the PL between 1922 and 1941, Mr. P.O. Pedersen in "Den Polytekniske Læreanstalts Udvidelsesplaner", *Ingeniøren*, 1932, pp. 113-28.
5. List of candidates collected by the inspector of the PL, A.N. Ørsted, c. 1879, published in A. Steen op. cit. pp. 183-207.
6. For a brief outline of this see G. Nørregaard, *Entrepenørforeningen, 1892-1942*, Copenhagen, 1942, pp. 22-32.
7. The statutes are cited in full in Johs. Kristensen, *Dansk Ingeniørforening gennem 50 Aar, 1892-1942*, 1942, p. 177. In connection with its 100th anniversary, the DIF has just published an account of its first hundred years, T. Morsing, *De Ansete Mænds Fagforening*, Teknisk Forlag, Copenhagen, 1992. (The Association of Respectable Men). For the early period this book does not add much to our present knowledge about the DIF; this is partly because most of the early archives were lost in a fire during World War II, partly because of the approach.
8. See Lundbye's account in Lundbye op. cit. pp. 227-244 and the accounts in *Ingeniøren*, 1892-94.
9. *Ingeniøren*, 1892, p.73 ff., p. 77 ff., pp. 81-85, p. 98 ff., p. 100 ff.
10. During these years we find quite a few accounts and discussions of the conditions for working abroad, especially in USA. See e.g. *Ingeniøren*, 1911, p. 109, L.P. Lauritsen, "Brev fra Amerika", (A letter from America). See also the recollections of the electrical engineer, cand. polyt. Asbjørn Winther, who worked in Edison's labs during the 1920's in *El-Nyt*, 3/1991, pp. 4-5.

11. A.Lütken, Beretning om Ingeniørundervisningen i Nordamerika, (An Account of Engineering Education in North America), *Ingeniøren*, 1894, pp. 165-67.
12. The new plan was cited in full length in *Ingeniøren*, 1894, pp. 95-110, 229-32, 237-8, 243-46, 252-56.
13. Aage Hannover (ed.), *Dansk Civilingeniørstat 1942*, Copenhagen, 1942, Oversigt over de fra den Polytekniske Læreanstalt udgaaede kandidater.
14. Absalon Larsen, "Den Polytekniske Læreanstalts Elektrotekniske Laboratorium 1903-1953", *Elektroteknikeren*, 49, 1953, pp. 451-57 and Lundbye op. cit. pp. 268-280. A more technical and detailed description can be found in *Den Polytekniske Læreanstalt, samlinger, laboratorier o. lign.*, Copenhagen, 1910.
15. Cited in *Ingeniøren*, 1901, nr. 3, pp. 9-23.
16. Lundbye op. cit. pp. 280-286, 311-348 and Rigsarkivet, Polyteknisk Læreanstalt, B Byggesager 800-816, 1913-50.
17. *Odense Teknikum – Teknikumuddannelsen 75 år*, edited by Odense Teknikum og Ingeniør-Sammenslutningen, 1980.
18. I.B. Hansen (ed.), *Polyteknisk Læreanstalt – Danmarks Tekniske Højskole*, 1979, pp. 47-49.
19. In *Ingeniøren*, 25. II. 1933, the director of the PL, Mr. P.O. Pedersen, stated that the PL hitherto had been handicapped in comparison with other technical universities because of the very rigid curricula and exams.
20. P.O. Pedersen, *Den danske Civilingeniøruddannelse*, a paper read to Industriforeningen 1/3 1937, Copenhagen, 1937.
21. A detailed description can be found in the autobiography of Jørgen Saxild, *En dansk Ingeniørs Erindringer*, Lindhardt & Ringhof, Copenhagen, 1971. (*Recollections of a Danish Engineer*).

Farmers and Agrarian Industries in the 19th Century: Who's the Boss?

Martyn Bakker

The fact that some people or groups of people do not readily adopt the biggest and most advanced technology available, whereas others do, has more than once caused bewilderment on the part of outsiders. Contemporaries have for instance poured scorn on late nineteenth century British entrepreneurs for being almost maliciously slow in copying foreign technology. These remarks have been repeated, unmitigated, by others – economic historians, economists and others. In a way they have even become part of the arsenal of "common sense" ideas which the general public, including politicians, holds on technological development.

The same kind of derisory descriptions we may find in historical accounts of other countries: the French have their early 19th century Normandy textile industry, and in the Netherlands entrepreneurs of the first half of the nineteenth century seem to have been possessed with a universal mental sloth where recognition of the obsolescence of their technology was concerned.

Now, industrialists and capitalists are, as a rule, condemned when they do not innovate, so there is an implicit idea that normal industrialists' behaviour is aimed at incessant innovation. There is another group of entrepreneurs, however, whose technological conservatism seems to be taken for granted. Those are the farmers, who made up for more than half of the population in most nineteenth century European countries. Historians and contemporary townspeople seem only struck with perplexity when these illiterate, coarse, dialect speaking people go for a new technology. For once they do so, they seem to do it all and at an amazing speed. Historical explanation then is contended with showing the perfectly sound economics that underlie the decision to innovate, for it can be proved easily that the new thing is far superior to its predecessor.

Now I am by no means going to challenge the economic evidence shown in this respect. I just would like to say that it's a pity that the explanation stops just there. For when there are obvious signs of a massive enthusiasm among farmers for something new, something foreign, we ought to ask two further questions:

- have we in fact been right in attributing a special "mentality" to the farming community, or is it really a caricature which has been fostered by townspeople ever since the Middle Ages?
- and secondly, if we allow for various groups to have their own ideas regarding change and the society they live in, wouldn't any further research into agrarian innovation be an excellent opportunity to show its extent and its workings?

I agree in general with those historians and others who say that attitudes towards change are by no means universally the same; who say that innovation, too, is a social process in which values are at stake. Now in the case of farmers and innovation, we seem to be dealing with a group which in its common attitude towards change does not seem to act to the textbook and laws of modern economics like puppets on a string. The question is, however, to what extent it is only farmers who are different from others in this respect. That is something which cannot be answered easily. What I will try, though, in this paper, is to show you some examples where farmers and industrialists alike took their decisions to change partly on the grounds of calculation, which one could call rational, partly in accordance with arguments that ultimately defy quantification and which I would call cultural.

Sugar

In 1858 three Amsterdam industrialists set up the first beet sugar factory in the Netherlands. It was not really the very first attempt to produce sugar from beetroots in Dutch history: in the days that the Netherlands formed part of Napoleon's Empire, the Emperor had ordered the country to grow thousands of acres with this new crop – but this was a shortlived experiment and, at least in the Netherlands, it met with no success for a variety of reasons. In France and Germany, however, some stubborn and rich men continued the line of thought set out by Napoleon even when the emperor spent his days on St Helen's. And so it happened that in the late 1830s a technology began to develop that produced beetsugar of such quality and at such costs that it seemed able to compete with the common sugar of those days, i.e. tropical cane sugar. Little by little this technology was elaborated further. Technicians, inventors, scientists and entrepreneurs agreed more and more on the issue of speed and scale of production. Making sugar is a combination of various chemical and physical processes in which the time factor is important. To this must be added the fact that a sugar beet must be processed very soon after it has been lifted from the field, lest it loses much of the sugar it contains.

Recognition of these time factors made that the ideas of the 1820s and 1830s were abandoned. Initially it was propagated that beet sugar ought to be produced in small cheap workshops annex to small and medium-sized farms. The farmer who grew beet, would be the manufacturer of sugar as well. This, however, became an almost impossible option. The technology of the 1840s and later demanded large amounts of capital to be invested in complex, steam heated and steam driven installations; apart from the financial aspect, this technology required a kind of expertise on the part of the manufacturer which even the larger farmers did not possess.

At this point of technological development, the Netherlands joined the ranks of beet sugar producing countries. The factory built in 1858 by the Amsterdam industrialists could process 60 tons of beet daily. In order to make a profit, it needed to process some 7000-7500 tons every winter. To obtain this amount of raw material, the directors had to contract at least 250 hectares for the first year and 250 hectares for the second year – one should never grow beet on the same field during two consecutive years.

Somehow, the directors managed to convince a sufficiently large number of farmers who agreed to grow this new crop. During the first couple of years their factory made interesting profits which caused other people to invest in this new industry as well. And each factory needed the assistance of probably over two hundred farmers to grow a sufficient quantity of sugar beet.

As shown here, the number of factories grew rapidly, and so did the surface covered with beet.

In fact this development was by no means obvious, for at the time there were important arguments for farmers not to go into sugar beet growing.

1. in the early 1860s they did not need any new crop at all. Their ordinary crops were profitable and there was no reason to see this situation come to an end;
2. sugar beet had a bad reputation: it was said that it exhausted the soil. It is a deep-rooting plant, and the idea was that all sugar contained in it by the end of the summer must have been drawn from the soil. These ideas were held by a number of leading persons in farmers' organisations in Holland, France and elsewhere.
3. In part they were right: a sugar beet is a demanding plant, and the danger of exhausting the soil is not wholly imaginary. To prevent this, and to obtain a good crop, in quantity and in quality, the soil needs to be ploughed deeply, more deeply than usual; sowing must be done very precisely and during

spring and summer the field must be weeded often. This requires much labour. Extra costs must be added for the purchase of a large quantity of fertilizer. A sugar beet needs much fertilizer, much more than had ever been put on Dutch fields, and in fact more than was readily available on the farms. Hence the farmers needed money to buy fertilizer.

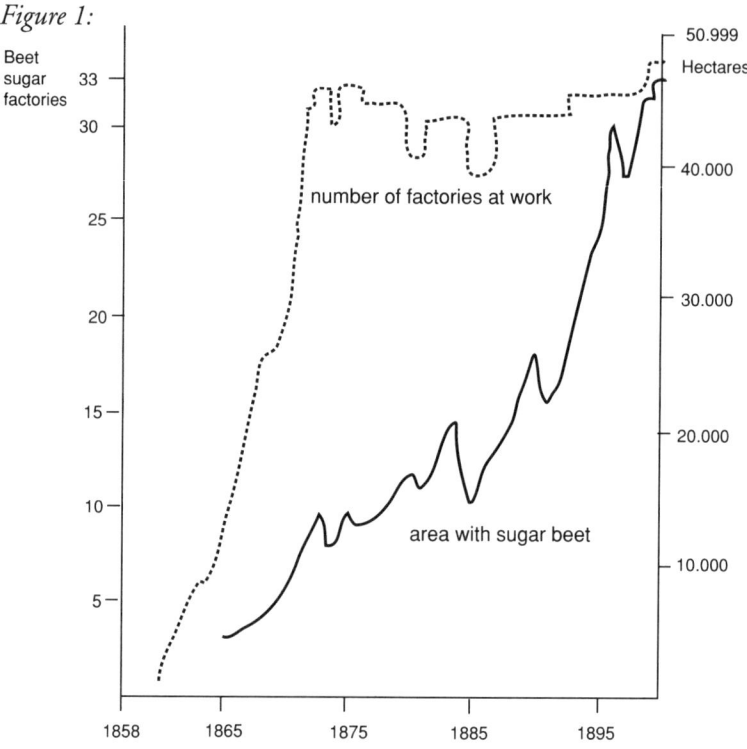

Figure 1:

So we have three reasons for not going into growing the new plant. Nevertheless, the new industrialists never seemed at a loss to find farmers prepared to sign a contract. Why? For the simple reason that these contracts were extremely profitable to the farmers. In fact, they were almost spoilt by the manufacturers. As soon as they signed a contract for growing some fields of sugar beet, the farmers were given an advance to buy fertilizer. Often, the factory even paid for the ploughing and the weeding. In all, the price a factory paid for one ton of beet was high, higher than any other crop seemed to yield per acre.

As soon as the number of factories began to expand, the farmers realised that they might try to put up prices even more, and they tried to stir up competition among factories over the price of beet. Several times, farmers tried to organise themselves in order to control prices. In this, however they did not succeed. The factory owners reacted and set up an association with regular meetings to discuss the situation along the beet front, and with the explicit aim to control prices. Understandably, it was much easier for 30 directors to stand united than for hundreds and hundreds of farmers, spread out over a large area and lacking precise insight in what went on in other regions.

The farmers tried in the mean time to gain even more money in other ways – illegal ways. They sold crops twice, they used seed of types of beet that produced enormous weights per acre, but contained only very little sugar, they economised on the use of fertilizer, etc.

Whether we should call these signs of a rising discontent, or merely excessive greediness, I wouldn't know, although I am inclined for the 1860s and early 1870s to call it greediness indeed, judging from a variety of archival sources and other evidence.

By this not very pleasant behaviour, at least a number of farmers sharpened the contrast between the interests of agriculture and industry, and the factory owners were confirmed in their probably hidden suspicions concerning the farmers' state of mind.

In the late 1870s, the situation began to deteriorate and in the 1880s it came close to disaster. This holds for both parties.

Sugar beet grow best on clay soil, which also is suitable for growing wheat. Grain prices began to fall sharply after 1875, and some other crops, like flax and madder, also saw a rapidly deminishing demand. Farmers began to offer their land for sugar beet spontaneously. This seemed a welcome opportunity for the factories to reduce prices, and within some years, the roles began to change: it was the farmers who became dependent on the willingness of industry. The sugar industry, however, was not seeking to reduce beet prices simply because they wanted to plunder the farmers. Raw materials made up some 75% of their total costs, so a reduction on that side would greatly affect the margin of profit. This margin became much smaller from the late 1870s onwards, since the world sugar market tended to decline. It collapsed completely in 1883, so the factories needed to take every measure to reduce costs. It turned out that no famer would sell his beet under HFL. 9 per ton, so there was an obvious limit. The other possibility for industry was to demand a high quality of beet to be grown, a beet that contained much sugar. There were various types of beet containing a high percentage of sugar, but these were very disadvantageous to the farmers, since one hectare produced only some 25 tons. These were small beet that needed to be planted at relatively large distances from each other, so one hectare would hold only a limited number of plants. Here again we see a clash of interests between farmers and industry – but we should bear in mind that it was still industry that had the money in spite of the recession – and it was the farmers who desperately needed it.

There are various dramatic descriptions of factory agents walking through villages with bundles of money and contracts loosely tucked under their arm, arrogantly handing out cigars to unwilling farmers, who knew they were mere puppets on a string pulled by almighty and unknown industrialists. The farmers needed the money given in advance by the factories to buy fertilizer; but they needed the money first to pay existing debts, and then had to make other debts to buy the fertilizer the contract stipulated.

On the other hand, very few factory owners managed to explain to the farmers why they, too, were in a difficult financial position. In fact, there are signs that they did not care to do so at all. They were fed up with cheating and complaining farmers, whom they hardly knew. Although quite a number of the directors originated from the small towns in the region, they considered themselves as different from the farmers, whereas the farmers never had felt any real sympathy for "them townsfolk". Each belonged to his own world, and only by chance their mutual interests coincided.

Butter

Let us now go from arable farming to cattle breeding, to dairying, to be more precise. For ages the Northern province of Friesland had been the Dutch dairy region par excellence. In the 1870s the Frisian dairy farmers found it almost impossible to compete with Denmark on the English butter market. Thousands of tons of butter had suddenly begun to pour of Denmark and it was of a quality far superior to what was being produced in Friesland. After various less successful attempts to innovate, the Frisians came across the butter separator built by Laval. Laval was a Swede who, together with the German Lefeldt, patented a separator in 1879.

This was the outcome of some 15 years of trying and tinkering, of more or less, but mostly less successful prototypes and small series. But suddenly they had a separator that was not expensive and indeed matched all requirements set out by potential users. Within a few years Laval produced various types of separators. The simplest one was hand operated, a slightly larger one could be driven by a poney or a horse, and then there was the largest model, which was often sold in pairs or more, with a 3 to 6 hp steam engine attached to them.

The separator turned out to be in line with the possibilities in Friesland to improve its dairy quality. At first, some individual farmers bought one, and their examples did not go unnoticed, on the contrary: they invited everybody to come and see. A few years later, there were some people from the butter trade who bought a steam driven separator and set up a network of milk-carts to collect milk at neighbouring villages. They bought milk at fixed liter prices and the farmers seemed rather happy with this new system which gave them an acceptable amount of money, for which they didn't have to do more than milk their cattle and fill the brass milkchurns. But as soon as these steam factories had started, quarrels sprang up. Farmers said that they had filled the milkchurns to the brim, which was 40 litres. The factory, however, said it had received only 38 litres or even less and paid accordingly. In fact, the farmers were right, in more than one instance.

The result was simple. In 1887, eight years after the first steam dairy factory had started in Friesland, a cooperative society was formed, the first of its kind in the country. 23 discontented and angry farmers with over 700 cows brought together enough money to set up a steam driven factory.

What happened next can be seen from this table (p.182):

By the turn of the century, buttermaking had become almost exclusively a factory based activity.

At the start some doubts had existed concerning the quality of butter made by a separator, but the British market turned out to react extremely well. The gospel spread from one village to the next; one cooperative society sprang up next to the other, know how was borrowed from each other and the province was pervaded by a "new spirit" – in which there was a common enemy: the traders and other people who set up non-cooperative dairy factories.

However, there were also some aspects to be considered very seriously before embarking on such a steam driven adventure. Especially the scale of operations gave rise to serious discussions. It was obvious that a large number of smaller farms could benefit from a factory with a separator driven by a 3 HP steam engine. Daily some 2000 litres of milk could be processed, but of course, it was said, "such a small factory wouldn't have room for a director with a large salary, a manager or a clerk or book keeper, an engineer or stoker to idle around."

Overhead costs, including the costs of staff which did not visibly take part in the actual production, were considered a serious waste of money. This was why there was so much opposition to privately owned factories: every penny that was paid to such persons – who were not farmers – was stolen from the farmer who got less for his milk. Therefore, the size of factories remained relatively small, so that it could be run completely by the local community.

If we compare the developments in Friesland with those in the south of the country, we come across some astonishing differences. Just take a look at this map:

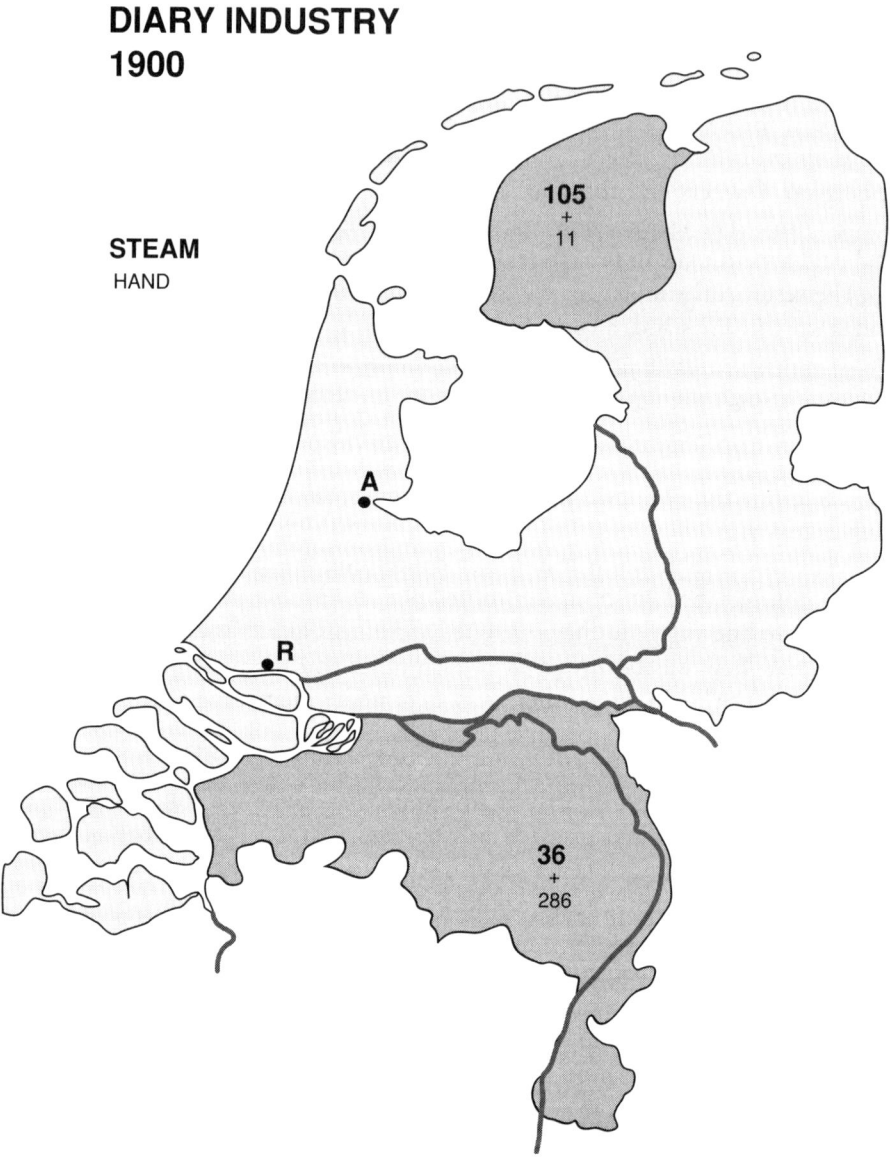

The South was later than the North. Was there perhaps no need to change, because butter making was of a higher standard? No, by no means. Dairying took place on an extremely small scale and the product was of such poor quality, that there existed hardly any trade. The small amount that was thought fit for export, was second, third rate butter that never met competition from Denmark. In the South, butter was very often still a means of exchange for farmers in the 1890s. They had just one or two skinny cows on their farms, standing in a shed for most of the year eating turnips and hay and producing dung.

All of a sudden, things began to change. The winter of 1890/91 had been very cold, and in the spring of 1891 a violent storm destroyed most of the crops near the hamlet of Tungelroy. The local schoolmaster, seeing the extremely depressed state of the farmers and their families, decided that they should try and follow the example of a village just across the Belgian border. One year earlier, some farmers there had set up a small dairy factory, which was said to be profitable. The schoolmaster managed to get the support of some 30 farmers, with 184 cows, and in May 1892 a very humble dairy factory produced butter in the village of Tungelroy. The word was spread by the newspapers in the area, and the following year at least 13 other similar factories were set up in the neighbourhood.

The factories were all cooperations, and they used the simplest of all Laval separators – the hand driven model. The costs of such a factory were low: some HFL 600 for a daily capacity of 1500 litres.

It is easy to understand why the farmers in the South chose for this small scale type of production. A factory needed to have the milk of cows standing not further than say two hours away from the factory. The local infrastucture in the South was bad. Farmers often had just a dog cart, or an extremely slow oxen. In two hours not much distance could be covered, so the potential milk area around the factory was not large. And within this already limited area, there were only very few cows, each of which gave less milk than its Frisian counterpart. So a steamdriven factory would probably have an immense overcapacity and hence the cheap, small scale hand driven technology was a very sensible alternative.

But there is more to the story. Here in the South, too, the preference for the farmers' cooperative society is evident. Here, too, non-farmers are considered as unwanted strangers. Especially the dishonest, or at least doubtful, activities of butter tradesmen and shopkeepers had for years or even generations caused anger and frustrations among the farmers. Especially the truck system was one of the evils which the cooperative farmers hoped to get rid of. But in some regions, people had their doubts as to the consequences of this new type of buttermaking. It stimulated farmers to have more cows and to produce more milk. The total quantity of butter produced would more and more excede the needs of the region, which meant that, in fact, the farmers became linked to distant markets, where untrustworthy merchants were merciless and would care only for the farmer's money.

The danger of "foreign" influence would become even more imminent in the case of steam factories. Then the local community would have to ask people from outside to manage the factory and its technology – a director, an engineer, a real bookkeeper. These were professions that were not available in the villages. In the small hand driven factories, it was the schoolmaster or one of the farmers' sons who had gone to school who could do the book keeping; the local blacksmith could provide spare parts, although this technology was so simple that there was hardly any risk of things breaking down.

Especially the South of the country by the end of the nineteenth century saw an overwhelming emancipation movement among the Roman Catholics, who at least until 1853 had officially been regarded as second rate citizens by the calvinists in the North. The South was Roman Catholic for 95%, and when we look at the leading characters in the cooperative movement, we see that there is also a number of clergymen involved. The cooperative societies were almost without exception named after a saint, often a statue or a cross was put outside the building.

The "foreign" influences mentioned above, not only refer to traders, and townspeople, but also to people of non-Catholic faith. In order to keep the co-operative movement in the dairy sector in line with the general emancipation of the catholic countryside, the propagators set up a whole system in which the community became self supporting: hundreds of locally managed hand driven dairy factories. They set up an extremely well organised trading network through which the small factories could sell their butter even abroad without the danger of being conned by foreigners; they set up an equally well organised training system for farmers' sons in which the latest developments in dairying were taught together with the most basic principles of calculus, mechanics, book keeping and farming in general. The basis of the entire organisation remained the parish. In this context, it becomes more understandable why in one village there could exist four mini-factories, where everybody came to bring his own milk, instead of one large factory shared between two villages and employing three men with horses and carts to collect the milk at the farms.

Science and Values

We saw the controversy between farmers and industrialists in beet sugar manufacturing, which was a struggle over the price and quality of beet. The conflicts in Friesland leading to the cooperative movement in dairying were also over the price industry wanted to pay: just a small sign of dishonesty on the part of the industrialists, and off went the farmers. As we have seen, setting up a dairy factory did not require heavy outlays of capital, so it was not so difficult for a group of farmers to go their own way.

The situation in beet sugar was entirely different. There was but one technology available, and there was, indeed but one technology that was profitable: a large scale, expensive production process. During the 1880s and 1890s the world sugar market was so bad that it would be wholly senseless to start a new factory on a cooperative basis. Besides, the general agrarian crisis of the 1880s had reduced financial reserves among farmers to virtually nil. There seemed to be one solution to mitigate the conflicts between farmers and industry, viz. to pay the farmer according to the quality of his beet. This had been subject of discussion and research in France and Germany in the1870s already and it was considered something from which both parties could benefit. The factories would receive a raw material from which they could produce more sugar without higher costs of production and in exchange for this, they would be able to pay the farmers slightly more – which would stimulate the farmers to take more care of his crops, spend more money on fertilizers, from which, in turn, other crops would benefit as well.

On paper this sounded all very reasonable. However, a sugar beet is a complex plant. It contains sugar but also a host of other chemical elements – and all of these have to be thrown out during the processing. But some are difficult to get rid of an can seriously obstruct cristallisation – which is the last phase in sugar making. So what would you call a "good beet"? The manufacturers said: not just one which contains much sugar, for when it contains many salts at the same time, it is almost as useless as a poor beet. By the 1880s there existed various simple ways of measuring the percentage of sugar in a beet, although their precision was not extreme. For analysing the quality and quantity of other elements, especially the least desirable ones, the methods were less practicable. This was stressed by manufacturers as an argument why they would not pay according to quality. It was impossible to analyse hundreds of samples a day, doing various analyses on each sample. But even when they would agree on paying more, just for the higher percentage of sugar, another problem arose.

Who would do the analysis, and where would the sample be taken? The one to do the analysis would, of course, said the factory owner, be one of my men, for I'm the one who is paying. And I want the sample to be taken at the factory, at the moment when the farmer delivers his beet. No, said the farmers, we don't trust your people or your laboratory equipment. We want someone neutral to do it – although we probably can't pay such a person or institute. In that case we will allow your men to do the analysis, but the sample has to be taken immediately before the beet is lifted from the fields. No way, said the directors, for it takes several days for the beet to be transported from the fields to my factory, and it may freeze and what not, and in the mean time the sugar level goes down very quickly, and I won't pay for sugar lost before arriving at my factory.

And then there was of course always the nature of the sample itself: how many beet per cartload, who was to pick them, and who was to control the controllers?

In short, discussions of this kind went on for years in various countries without any solution that found general approval.

However, there was one unorthodox Dutch factory owner, J.F. Vlekke, who owned two factories and who sincerely thought that most of his colleagues were selfish and talked as much nonsense as most of the farmers. The important thing to do, he said, is to go to the farmers in person and present a plan and discuss it with each of them in private. His plan was ready in 1895, and it is indeed a brilliant and tactical compromise.

He began to say that, indeed, the sugar percentage is but one of many important variables that make up the value of a beet. But by that time there was an extremely simple and especially fast way of determining it with some exactness. For technical reasons, the analysis should take place at the factory and the samples should be taken at the gate of the factory. Vlekke would build a special laboratory in which hundreds of samples could be analysed every day, and if farmers did not trust the results of the analysis, they could ask for an outsider's second opinion, the costs of which would be paid by the factory. It would be up to the farmers whether they delivered their beet for the normal price per ton, or according to the sugar percentage. In the latter case, the farmer would receive the normal price per ton as a minimum, and to that would be added an extra sum according to its higher sugar quality. Vlekke went even so far that the farmers who chose for the quality contract, shared in profits of the factory.

The plan, as I said, looked good on paper. And then Vlekke went round in person to visit the 1.300 farmers who normally grew beet for his factories. Just over three hundred responded favourably at first sight. The result was that Vlekke received a large amount of high quality beet, and he was able to pay those farmers a better price. The second year the number of participating farmers more than doubled, and the quality of beet steadily rose.

To Vlekke's colleagues, this way of contracting beet was close to a dangerous kind of socialism, cooperation with farmers was considered something close to collaboration with the enemy. The cooperative movement among farmers in general was despised as something evil.

Instead of noticing how Vlekke had lost the need to take part in strenuous and nerve wrecking fights over beet contracts and prices and could devote his attention to the factory, the other directors continued their war – among themselves and between them and the farmers. There were only two other factories to follow Vlekke's contract: one new factory in the extreme North and one near Amsterdam – both far from the heartland of sugar production in the South-West.

It was in 1908, shortly after Vlekke's death, that rumours spread saying that the shareholders and the new management of his factories wanted to get rid of this famous contract. Within a couple of months, many farmers decided to go their own way and set up a giant cooperative beet sugar factory, more or less at the doorstep of Vlekke's factory. It was the second of its kind, the first one having started production in 1899, in a neighbouring province.

Vlekke's contract produced very good results, and he was the first one to bring them into the open. But the amazing thing is that none of his colleagues even dreamt of following his unorthodox line of thought. Archival sources show many signs of the idea that there was an almost naturally determined enmity between both interest groups, farmers and industry, and that this law of nature must not be violated. We have seen it from both sides now, from the farmers' side in the dairy industry and from the industrialists' side in the sugar factories.

In either case, one could do sums and show that another solution would be more profitable: large scale production, both in Friesland and in Brabant could at an early stage already be more profitable than the system of one or more factories in each village. This is what we see with the non-cooperative dairy factories, which – in spite of the massive cooperative movement – continued to exist. As a rule they are larger than the local cooperations. From the farmers' viewpoint, this may be equally amazing, since most of the cooperative societies were able to pay more for the milk than the privately owned factories were willing to pay.

But in dairying as in sugar making, there also were farmers who as a matter of principle did not want to be involved in cooperative societies at all. Therefore, the rise of cooperative sugar factories in the early 20th century did not mean that the other factories came to lack sugar beet.

It just goes to show that it is only in part a matter of bookkeeping and numbers. It is very much a matter of "us" and "them", and when "they" come with a calculation showing that it is good for "both of us", you better beware. It depends on who does the sums that determines the outcome of a calculation. And it equally depends on the attitude towards the other, whether he is regarded as an opponent or a partner, if some uncertainty is taken for granted or used as an argument against change. We saw that in the sugar industry both parties said that they could never trust the other, and each said that scientific solutions were only in favour of the other's interest. But as soon as Vlekke was accepted by the farmers as an honest man, the relative roughness of the scientific method he used, and which he openly recognised, ceased to be an obstacle.

The same can be seen in the dairy industry: it was possible to determine the quality of milk in terms of fat quite easily by 1890. This would be the best method to pay each farmer according to the quality of his milk and it would immediately betray those who had added water to the milk. But the cooperative societies hardly showed any interest in this method, which – it seems – was used extensively in Danish factories, cooperatives and others. But both in Friesland and in the South, the members of the cooperatives trusted each other and didn't consider it necessary to grab every half penny immediately. It was more important to have the general sense of harmony and unity within the local society, a feeling generally provoked by the greediness of "the others", the outsiders.

In my introduction, I suggested that people undeniably have used economic, quantifiable, arguments to decide whether or not to innovate, but that it would be interesting to try and have an open mind to other arguments as well. The previous examples I think have shown us that it is not only farmers who combined economics and cultural aspects and personal arguments to account for their use of technology. Industrialists themselves, although they claim to be different, are equally "human" in their acceptance or rejection of new pieces of technology, scientific knowledge and the like. In fact, it shows us that technologies are judged according to the extent to which they fit in with the values held by its potential users, and those values are by no means restricted to economic values.

Note:

This paper is entirely based on two books and the sources referred to in those books, viz.:

- M. Bakker, "Ondernemerschap en vernieuwing. De Nederlandse bietsuikerindustrie, 1858-1919", Amsterdam 1989 [Entrepreneurship and innovation. Dutch beetsugar industry 1858-1919]
- M. Bakker, "Boterbereiding in de late negentiende eeuw", Zutphen 1992 [Dairy making in the late 19th century]

For additional information, please contact the author through:

Eindhoven University of Technology
Faculty of Philosophy and Social Sciences
PO BOX 513
5600 MB Eindhoven
The Netherlands

Danish Modernization Strategies – from above or below?

Dan Ch. Christensen

I.

It is the aim of this paper to explain how Denmark entering the 19th century so dependent upon technology transfer from abroad could emerge at the turn of the 20th century as an innovator in agricultural technology and organisation and set the model for a range of its neighbours.

An historical explanation of this development requires an analysis of the reception and diffusion of technologies transferred from Great Britain, mainly the cultivation system of convertible husbandry and alternating crops on the one hand and agricultural implements as the swingplough with a cast mouldboard, the seed-drill and the threshing machine on the other. I shall try to show that the agricultural improvements fit nicely into the estate economy, but rather poorly so into the peasant economy. The cases will illustrate the outcome of modernization strategies from above.

Secondly I shall try to show how technological institutions of education and R&D tended to mirror a similar socio-economic conflict in society, i.e. a strategy of modernization from above under the leadership of the aristocracy and the university and polytechnic of Copenhagen (possessors of capital and knowledge), as opposed to a development strategy from below to be carried out by cooperation between peasant organisations and folk high schools (pooling their capital and knowledge). These opposing strategies were intimately related to the question of technological control and the main adversaries, H.C. Ørsted and N.F.S. Grundtvig were both deeply influenced by foreign experience.

Finally I shall attempt an assessment of the Danish cooperative dairy farming. Why was it so successful as a model that it not only diffused nationwide very rapidly, but was also transferred to a wide range of developing countries? This dairy farming system implied a rural process of industrialization, but one should not ignore the reverse side of the medal which was probably a delay of an urban industrial take-off. Both features are characteristic of the peculiarity of the Danish modernization process. The cooperative movement is the conspicuous example illustrating a modernization strategy from below. I hope to demonstrate through these cases that modernization is not a simple matter of technology transfer as such but of adopting a specific technology favouring a predominant cultural pattern.

II.

The historical background of the Danish modernization process was a range of land reforms during the last quarter of the 18th century. The most important elements were the conversion of the communal open field system, the incipient abolition of villainage and the gradual enclosure and private ownership of farm land. Although progressive landowners here as in the UK took advantage of these reforms towards a freer market economy by higher investment in technology and labour, the copyholding peasantry in this country was protected as a class by the absolute state by legislation inhibiting the amalgamation or parcelling out of the existing 60.000 farms. Consequently there was no way the aristocracy could buy up land or lease it to farmers. As opposed to the UK nine tenths of Danish agriculture was in the hands of a stable middle class peasantry, and there was no urban pull to attract a rural proletariat.

Historians of these land reforms have generally focused on state incentive to humanize the peasantry by means of human rights, education and individual economic responsibility. There have also been interpretations emphasizing the fiscal interest of the state. Technological modernization from below has rarely been taken into consideration. Innovation in terms of crop rotation, convertible husbandry, iron cast implements, etc. were harbingers of a more rational cultivation system, the blessings of which were acclaimed by Danish aristocrats. The immense public debate was mainly arguing for or against the agricultural revolution in the UK. As an historian of technology I'll try to understand the character of the profound transformation of Danish society in the 19th century by asking the historian's traditional cui-bono?-question, rather than accepting the rather naive idea that the absolute state should have cared mainly for the human rights and social welfare of the peasantry.

III.

The traditional 3 course rotation system (rye, barley, fallow) was exhausting the soil. The communal village system was trapped in a vicious circle of diminishing returns. The only way out was a radical change of the cultivation system. From the point of view of the peasant indefinite villeinage was a gargantuan school of indolence. From the point of view of the landowner it was even worse since villeinage implied a technological dependency, because the tilling of his land suffered from the use of the traditional implements and routines of the villeins. Rational cultivation systems demanded modern implements and hence hired labour. This was a costly and risky investment. Consequently the estateowner would take active interest in his former copyholder's ability to pay mortgage and other fees converted from services in kind, like tithes and villeinage. In other words the more productive the cultivation system of the peasantry the better for all parties concerned. Technological interdependence between manors and peasants was an obstacle to modernization.

The 3-course rotation system had fulfilled the needs of subsistence for centuries. A shift in favour of crop rotation including clover, pulses and potatoes claimed higher investment in implements, fencing, animal stock, and even if scientific arguments in favour of a more rational system were convincing for the scientific mind, the peasant was hesitant to abandon traditional practice. With the abolition of the open field system dominated by strip-cropping, how should he enclose his fields and what should he grow? He was advised to grow more fodder and to feed his stock indoors during the summer to improve the fertilizing power of his soil and hence increase grain yields. But this was at best a long term project. So instead the average copyholder would take recourse to some variation of the traditional 3-course rotation system, excluding nitrogenous fodder crops, and resort to a maximum of grain fields interrupted by a minimum of successive grass land and fallow. And this certainly was not breaking the vicious circle of exhaustion.

Could copyholders be forced to modernize? If the scientifically minded landowners were keeping the recipe of progress, were they not justified to enforce its realisation? Physiocrats were promoting capital investment and Thaer and his disciples were fervent believers in chemical analyses and calculation. Pulses, turnips and potatoes would invest the soil with sustained power. Encouragement and premiums to copy-holders and free-holders adopting a crop rotation system were introduced by the Royal Board of Agriculture (1769). But would voluntary measures suffice?

The Royal Board of Agriculture gathering the most progressive estateowners were deliberating a legal enforcement of a planned cultivation system. Copyholders not complying with the regulations of their tenure – to adopt a system of crop rotation – ought to be liable to loose it. Lawsuits ended up with fines or loss of tenure, until the Provincial Consultative Assemblies during the 1830's took a new stand to protect the independence of the tenants. At the same time the theory behind the crop rotation system was proved insufficient. The perfect system based on a successive calculation of the nourishing and corroding forces appeared to be a phantasm rather than empirical soil science. When in the 1830's the Royal Society of the Sciences invited a prize paper on the advantages of intensified cultivations systems, it had to promote a new member for the jury. The choice was a die-hard believer in crop rotation, so papers arguing against the scientific ideal of crop rotation and indoors feeding of cows failed. The ultimate strategical goal of the Royal Board of Agriculture was to lift the ban against amalgamation of copyholds and introduce leasehold farming along British ideals. The result they imagined would be a polarization of former copyholders into rational farmers on the one hand and a landless labour force on the other. They did not succeed.

The abolute monarchy stuck to its protection laws in favour of the middle-class peasantry, whose practice ran contrary to inadequate theory. At least three reasons serve to explain the prevailing cultivation system. Wheat and barley were valuable export products, and the abolition of the British corn laws made them even more favourable. Secondly the poor stage of dairy and food processing technology made dairy farming unmarketable except for landowners. Thirdly the economic laws of Thünen explained why distance from the market place was a key factor in cost/ benefit calculation of extensive/ intensive cultivation systems. The average peasant simply could not afford the investment in new seeds and better tillage and larger stock. He could not pay mortgage and fees without the cash crops of as many rotations of grain as possible.

To sum up: the cultivation system transferred from the UK was developed by British landowners and suitable for their Danish peers. The peasantry lacked the capital and knowledge to adopt it and efforts to force them failed in the long run. The scientific theory behind it was premature and the traditional technologies of the peasantry survived with only minor changes. [1; 2]

IV.

The second element in the technology transfer from the UK were agricultural implements, such as swingploughs, horsehoes, seed drills, chaff-cutters and threshing machines. The advantage of these implements is commonly believed to consist in the saving of labour costs. Well, this is one of the valid arguments for buying a threshing machine, but the other implements were introduced as part and parcel of a more intensified cultivation system, that of crop rotation and convertible husbandry, which increased labour costs. These technical innovations mainly derive from the increasing use of pig iron for casting mouldboards and a gradual substitution of wood by iron in agricultural implements. An important improvement in plough construction was the alledged scientific mouldboard. Bailey and Jefferson quarelled about the right of paternity of this invention. Both claimed to have used mathematical calculation in optimizing the curves of the mouldboard, but this claim was soon disputed. Ransome, the plough manufacturer, regarded it as a typical enlightenment attempt to give an invention a status of authority.

It is not so complicated to trace the stages of this technology transfer to Denmark. Firstly progressive landowners ordered these new implements from England. Secondly skilled journeymen and iron founders of British origin immigrated in order to produce these implements, attracted by the fact that neither patent fees nor competition would cause inconvenience for them in this country. Thirdly, a generation later, Danish mechanics began imitating British proto-types or producing their own innovations. This was almost a start from scratch, since the Danish-Norwegian twin kingdom due to natural endowments had established a virtual division of labour, where the Danes exchanged corn for Norwegian iron products. Still, around 1825, modern British implements had diffused almost exclusively among landowners realising that modernizing their cultivation system involved wage labour and investment in new technology to the detriment of villeinage services. For a peasant to acquire this new technology was simply out of the question. For economic as well as for technological reasons.

A complete swingplough with a cast iron mouldboard was very expensive. A copy-holder or a yeoman in favour of a modern cultivation system had to rid himself of the burden of villeinage by having it converted to fees. To this must be added his mortgage obligations and one will understand that he would always be short of cash. Secondly, cast iron parts could not be repaired locally. Thirdly, the peasants of a parish would normally have contracted an agreement with a smith, to serve their technological needs both for tilling their own soil and for fulfilling their obligations as villeins. As a rule the smith was allotted a parcel of land and other natural means of subsistence, in return for which he was obliged to shoe horses and repair implements. The annual deliveries to the parish smith would be wasted if a peasant replaced local supply by modern technology of urban origin. For these reasons peasant economy would stick to traditional technology.

The Royal Board of Agriculture launched a virtual campaign to speed up modernization at the beginning of the 19th century. The general attitude was in favour of the British way of farming. Land reform was accused of paying too much attention to human rights and socio-economic interests of the peasantry and too little attention to rational agriculture. Since agricultural modernization depended on capital and knowledge, the aristocracy propagated by means of encouragement and persuation the introduction of modern implements. The English plough was given a symbolic value in the modernization process. Distribution of these ploughs was undertaken as were prize awards in plough competitions. Although there were excellent Danish manufacturers of cast iron mouldboard ploughs, copies of Bailey's swingplough were distributed, simply because of their symbolic value. And Danish improvements of the wheel plough were banned from plough competitions, irrespective of their high quality. Since only English ploughs were allowed, the emphasis of these competitions changed from selecting the best implement to awarding the most skilful ploughman serving a landowner.

As these measures did not produce the desired effect, the Royal Board of Agriculture made an attempt to get rid of the alledged incompetence of the village smith, by enhancing the qualifications necessary to obtain the right to carry out trade. This was a direct threat against the local technological expertice of the peasantry, which was lucky enough, however, that a majority of local officials defended the right of the peasantry to choose their suppliers themselves. So the aristocracy suffered another technological defeat.

The outcome of this controversy was typical and significant. The royal iron foundry of Frederiksværk, which had been established as a military-industrial complex, began a large scale civil production of iron cast mouldboards and spare parts, which were marketed by local merchants to peasants or village smiths and assembled locally often on the traditional wooden frame. In this way the improvement of farming implements took on a hybrid character of piece meal renewal.

During the 1840's more than 20.000 mouldboards a year were produced by two iron foundries, one of which was private, the other one, Frederiksværk, was beyond market competition. The division of labour between high tech and tradition hampered the urban iron foundry which only emerged on a large scale a generation later, when rising corn prices and falling iron prices and better communation systems made urban supplies, such as complete ploughs and threshing machines, more reasonable for the peasant economy. Still, however, the peasant community would maintain its contract with the parish smith.

The clash of interest was generated by the fact that transferred technology was indeed improving the productivity of 1/10th of the soil, but on the other hand it was a costly threat against the technological self-determination of the 9/10th. The rationality of the modernization strategy was not simply a question of technological artefacts or cultivation systems as such. Their rationale was embedded in the culture of large scale agro capitalist production. For the peasantry it was more rational to face the modernization strategy from above with a critical eye and make it conform to their own indigenous cultural environment.[3; 4]

V.

This clash of interest was accompanied by a parallel controversy about the function of technology in education. In fact we are facing the main ideological issue of this country during the 19th century casting its shade far into the 20th. I shall name this conflict between a public and scientific code of education on the one hand and a popular and layman's code on the other after their internationally well known leaders, Ørsted's line, the university-code, and Grundtvig's line, the high-school-code.

Ørsted won international fame for his discovery of electromagnetism in 1820; he was secretary of the Royal Society of Science, powerful professor at the University of Copenhagen, founder of the Society for the Diffusion of Scientific Knowledge (1823) and of the Polytechnic Institute (1829), and second to none in the Danish scientific community from 1815-1851. Grundtvig was a priest, a hymn-writer, an historian, a mythologist, an educational philosopher and a politician, in fact the only Dane who has caused a popular movement to be named after him and given basic inspiration to lasting features of Danish particularities, such as the folk high-schools and the cooperative movement. Whereas the Ørsted-line is elitist, idealistic and stresses the prerogative of the State in cultural and educational affairs as well as the hierarchical subordination of technology under science, the Grundtvig-line emphasizes private and popular grass-root initiative, the subservient role of science and technology to human ends, distrust of the State and bureaucracy, and it is hermeneutic and empirical. In short: development strategies from above and below.

Let us examine a little more closely the different positions of the two codes. When Ørsted was a young scientist working with galvanism and trying to gain recognition by the French National Institute, he was declining an offer to become the leader of a new technological institution. He professed a distaste for mere technology and took an oath on devoting his carreer to scientific theory. In a speech delivered in celebration of the Lutheran reformation he urged university students to devote themselves to science as if holding a divine service. Ørsted's model of society was Platonic. Science and philosophy constituted the head of organic society. Technology belonged to the stomach, the organ of the merchants. Science was autonomous, whereas technology invariably was mixed up with commercial greed. Science in Ørsted's worldview was worthy to be a cognitive end in inself and should never succumb to the temptation of degrading itself to becoming a means to something else's end. During the time when the crucial experiments on electromagnetism were on his mind he was witnessing a plough competition in his capacity as a member of the technical committee of the Royal Board of Agriculture, his outstanding contribution consisted in translating

Schiller's ode to the plough. Ørsted took great pleasure in thinking of himself also as a poet. He certainly kept technology at arm's length, except when technology would be exploited to harness his scientific goal.

After his discovery of electromagnetism he went on a grand tour of international celebration and upon his return he brought forward the idea of a Society for the Diffusion of Scientific Knowledge. This idea, according to Ørsted himself, came to him 'as a revelation' on board a ship from London. It should not be underestimated, however, that he had become acquainted with Sir Humphrey Davy, professor at the Royal Institution, London. Both scientists had by now been recognized by the Institut National in Paris. Both scientists were natural philosophers, which is apparent from Davy's 'Travels of Consolation' and Ørsted's 'The Spirit of Nature'. More relevant perhaps are the lessons to be drawn from Davy's experience as a professor of the Royal Insitute. In a nutshell they amounted to strain and frustration, because his employers at Albemarle Street insisted on useful research applicable to agricultural technology, like tanning, to the detriment of pure scientific experiments with galvanism. Davy's scientific knowledge had indeed produced some remarkable technological results, but the moral of the relationship between science and technology was conclusive: be sure that scientists are in control. A similar hierarchical order was the lesson taught by the French Minister of the Interior, Count Chaptal, in his text-book of chemistry. Give mechanics and industrialists an inch and they will take an ell of the life of a scientist!

Upon his return Ørsted became engaged in a struggle to put science on an equal footing with theology and the arts at the University of Copenhagen. The accomplishment of science since the Renaissance ought to justify a faculty of science. But Ørsted was miles away from a Baconian strategy. Science must never be degraded to a subservient role. To avoid this and to consolidate his position it was vital to control technological initiatives, keep the guild system in a subordinate position and educate scientists for governmental administration. Let me indicate some examples:

In 1823 The Society for the Diffusion of Scientific Knowledge was founded. The subscription fee made it unlikely that artisans would be able to afford to attend lectures. The overwhelming part of the audience was state officials and academics. Still a private foundation supported some artisans, but they were complaining about the abstract and futile character of the lectures. Ørsted responded that applicability and rules of thumb were not the business of scientists. The purpose of the lectures was to elevate the culture of the audience and stimulate its appetite for scientific knowledge, not to interact with the practice of craftsmen.

In 1828 a proposal to establish a higher education for technicians was submitted to the king. The idea, fostered by the editor of a succesful magazine of technology, was a balanced interaction between scientific theory and technological practice. The proposal was referred to the University of Copenhagen and Ørsted made a radical change of the programme by reducing the part of technology in favour of science. The ensuing Polytechnic Institute was an amalgamation of the would-be-faculty of science at the University and the Society for the Diffusion of Scientific Knowledge in terms of staff and funding. The new institution only made room for two technically competent members, Ursin, the abovementioned editor and professor of mathematics, and Winstrup, manufacturer of agricultural implements and the first Dane to have built a steam engine. Before long, however, both had resigned their posts due to Ørsted's negligence of technology. Science lectures left no slot for technical practice. This certainly was a rupture with the preceding position. Ørsted's predecessor as secretary of the Royal Society of Science had taken part in the interaction with mechanics, like Winstrup, and with technological institutions like the Naval yard and the Royal Board of Agriculture. There used to be much coming and going between lecture hall and workshop floor, to use the phraseology of T.S.Ashton. During Ørsted's reign there was almost complete separation and science was in complete control.

In 1842 Ørsted made similar comments on a proposal to establish an Agricultural Institute. He vigorously counteracted a curriculum integrating theory and practice. He saw no justification in founding a new institution aming at interaction between agricultural practice and scientific experiments. In stead he argued in favour of an extention of the would-be-faculty of science to include agricultural sciences as botany, zoology, geology, chemistry, etc. Practice was the business of the farms and dairies themselves and they belonged to a lower order in his Platonic hierarchy.

Although Ørsted's discovery of electromagnetism was the constituent principle of the telegraph, he never took any part in the development or introduction of this new technology. Similarly, his invention in 1826 of a method to produce aluminium remained unused, except for a set of plates considered more prestigious than gold plates by the royal court. During the first 25 years of its existence the Polytechnic Institute had graduated 87 candidates, 60 of whom made a career in the service of the State, only 16 within agricultural or industrial technology (11 died or had unknown carreers). In conclusion Ørsted was a very successful organizer who achieved to fullfil his goal, the establishment of a science faculty as a manifestation of the equal status of science vis a vis philosophy, theology and the arts. I am not trying to postulate that Ørsted actively counteracted technological innovation, because the examples of Søren Hjort and August Colding testify to the fact that he supported them. The point I have been trying to make is that Ørsted's code was elitist, hierarchical and controlling.[5; 6]

VI.

It is rather easy to compare Grundtvig's attitude towards the peasantry with Ørsted's code, because both wrote dialogues between a scientist and a peasant. Ørsted's scientist persuades his peasant to contribute to the erection of a monument of Tycho Brahe, because the common man ought to be grateful for the discoveries of science which will ultimately enlighten human life like a lighthouse. Ørsted's peasant has no ideas of his own, he just listens, gapes and conforms. Grundtvig's code reverses the roles. His peasant virtually examines the scientist and accepts no evasive answers. His dialogue takes place on a high school founded by peasants who also take the initiative to invite relevant scholars to serve the people. Consequently high school pupils do not hold professors in authoritarian reverence. Grundtvig exposed the view that by means of latin and mathematics universities educate students away from the people. His ideal peasant-student forces the attention of scholars towards relevant research, and to assess the quality of research the high school to be established in Sorø will install a touchstone of technology.

Behind this attitude lay Grundtvig's criticism of the enlightenment strategy, aiming at science based improvement from above, whereas genuine progress should be generated from below, taking off from 'the living word' of indigenous culture. Grundtvig's code of education was sharpened during his stay in England from 1829-31, where he witnessed the deprivation of control over peasant and artisan production by capitalist technology, reducing humans to slaves of machinery. The direct inspiration came from his acquaintance with the Mechanics' Institutes mushrooming during the 1820s. Exactly at this time they experienced a dispute about control. Were they to develop into employers' institutes or machanics' institutes? That was the crucial issue.

Although Denmark was an underdeveloped country in comparison with the UK from an industrial point of view, Grundtvig was convinced that technology would dominate universally in the future. Still he believed that ordinary people were endowed with sufficient intelligence to shape and control it, provided the peasantry was encouraged to gain individual and collective self-confidence. His task was to provide the inspiration for this self-confidence. The means to this end he found in the empirical evidence of a popular culture resisting uncritical technology transfer. He was mocking the academic establishment for its empty studies of latin and mathematics condusive to diplomas and offices only, and he encouraged peasants to form their own educational institutions in order to maintain and develop traditional values such as technological autonomy and control.

Translating Grundtvig's code into Habermasian terminology one could say that Grundtvig is protecting the life world of the peasantry against attempts of anonymous powers to systematize and rationalize household production and family life, thus reducing mankind from being an end in itself to becoming the means of external forces operating in the system world.

In 1855 when a second proposal to establish a University of Agriculture was being discussed by Parliament, Grundtvig fiercely protested against this State institution focusing on examinations. Agricultural education ought to be a matter for the farmers themselves and knowing that they were qualifying for the purpose of running their own farms they were in need of neither government funding nor diplomas. He also criticized the curriculum for wanting the core subject matter of history. As opposed to the ahistorical Enlightenment attitude Grundtvig was a typical Romanticist preoccupied with the inspiration and wisdom collected by past generations. 'Deed is the watchword of the Spirit!' in Grundtvig's code.

However, during this period peasant high schools were mushrooming allover the country. Some of them emphasized science instruction and borrowed scientific instruments from the Society for the Diffusion of Scientific Knowledge. Still it would be a crude exaggeration to suggest that they were primarily technological institutions. Their main purpose was to install individual self-confidence and cultural self-awareness. Control over technological development was seen as an important element to this end. [7]

VII.

So far we have been investigating modernization strategies from above and reactions from below. The transfer of agricultural technology from the UK carried ambiguous rationalities. The Ørsted code of scientific and technological education was suitable for scientific enlightenment and the formation of a technological bureaucracy, but too aloof to influence technology development in agriculture and industry.

During the 1870s a dramatic decline in the world grain market had occurred mainly due to huge American exports at very low prices thanks to low production and transport costs (cheap land, reapers, railways, steam ships). This caused a world-wide grain crisis, but it was, perhaps, particularly severe for a country like Denmark which had become so dependent upon grain exports to the UK. Something had to be done about it. The solution to the problem was to buy cheap grain and other foodstuffs and make dairy products. From 1882 to 1897 more than a thousand cooperative dairies were established. To understand the rise of the cooperative movement so successful in dairy farming that it was soon to be conceived as a pattern of technology transfer from Denmark, we must turn our attention to the traditional peasant mode of cooperation, and the Grundtvig code of technological innovation including the high school movement, grass-root initiative and the interaction between private iron foundries and machine industries and the research lab of the Agricultural University as the touchstone of viable technology.

The core elements of the cooperative dairies are the centrifuge and pasteurization. The centrifuge itself was based on an old well known principle. The utilization of this principle for the skimming af milk was brought about in a series of piece meal innovations, starting in Germany by Lefeldt and to be continued here by Winstrup, the son of the steam engine constructor who was removed from the Polytechnic Institute by Ørsted. Winstrup handed the results of his experiments over to Fjord, who was head of the research lab at the Agricultural University, and at the same time developing Pasteur's scientific discoveries in bacteriology to bear upon butter production. Fjord in turn established contact with a selfmade mechanic in Roskilde, Nielsen, who was employed by the Maglekilde Factory, being the private iron foundry that had produced a great many mouldboards for the peasant ploughs. Nielsen's contribution to the centrifuge was to make its operation continuous and more effective. The continuous centrifuge was a great leap forward, because it changed it into a machinery of economics of scale. Before Nielsen a centrifuge was an excellent machine in the hands of the estate dairy, because it processed 30-50 litres of milk. The continuous centrifuge separated several hundred litres of milk, and did so more effectively, thus reducing considerably the cost and energy as well as the quantity of milk necessary to make one lb. of butter.[8]

This is the technological element, but to make a cooperative movement the application of an organisational principle was required. This can be traced to the tradition of the collective peasant-parish smith contract relationship. As we have seen agricultural reforms had disrupted the communal principle and the Royal Board of Agriculture had made attempts to disrupt this local technological autonomy. The high school movement was aiming at bridging the gap which had arisen out of the individualizing trends of the enclosure movement. Now the technological cooperation was extended by the collective contract with a dairy manager, who must be qualified to coordinate milk supply and processing by means of highly specialized equipment. Cooperative dairies mushroomed all over the country. Its outspring was in western Jutland in 1882, almost as far from Copenhagen as possible. It diffused like a tide eastwards, and within 10 years there were more than 700. The production system multiplied and called for marketing coordination because the British export market excercised a quality control and a invited a price policy. As was the case with the plough technology, the cooperative dairy movement was a combination of local organisation and central high tech to which was added an overarching export organisation by the cooperative farmers themselves.

How did landowners, who had so far been the avantgarde in dairy farming, respond to this development? Either by applying the same technology principle through the working together of a handfull of neighbouring estates, or by joining the local cooperative dairy. In either way the estate dairies were engulfed by the cooperative movement. The radically new property of the continuous centrifuge was that by collecting milk in varying quantities each litre of milk obtained equal value, whereas earlier the small quantity from the crofter was insufficient to make good butter. By the traditional Dutch method of skimming the cream only estate dairies produced enough milk to make good butter for the export market. The continuous separator was particularly suitable to meet the demands of cooperative production, because equality was a principle built into the separator. So the estate dairy had lost its competitive advantage of scale. Similarly a principle of equality was built into the local cooperative dairy. Influence was accorded per person, i.e. the one man/ one vote-principle, irrespective of the member's number of cows, whereas economic return, of course, depended on quantity of supply. This democratic principle was on the agenda at high schools, being deeply involved in the national struggle for parliamentary democracy. 54% of the first generation of local dairy chairmen had been high school pupils.[9]

The continuous centrifuge did not make the peasantry dependent upon urban expertise, because a number of suppliers, e.g. Burmeister & Wain, which bought the Maglekilde patent, and the Swedish Laval Separator, which took the lead on the world market by buying a German patent, were queuing up in the villages to secure orders. The technological expertise was situated at the research lab of the Agricultural University, which was characterized by an interactive relationship between science and technological practice. One might say that it performed the role of the touchstone, which Grundtvig suggested as a key principle of the high school. The main part of construction costs came from local members of the cooperative societies. Statistics from 1897 show that rural food processing industry was by far the largest consumer of steam power, running 33% of the nation's steam engines. In terms of steampower urban industry was dwarfing vis a vis the rural food processing industry.[10]

When capitalist baconfactories approached the cooperative movement to offer some sort of joint venture, the cooperative societies declined for fear of coming under capitalist control. Distrust of modernization strategies from above and suspicion of allegedly rational large scale technology frightened the peasantry having been reminded of their historical experience at the folk high schools. Moreover, pig farming was becoming more profitable, because whey, the waste product of dairy farming, was returned to the farms as valuable fodder. The cooperative movement was certainly riding on a wave of success and was adamant not to risk their newly won independence.

The successful organization of the Danish food processing industry and its overwhelming export orientation gave very little stimulus to urban industry. Only in the 1960s did industrial export exceed agricultural export. Just 30 years ago!

VIII.

It has become almost a cliché to talk about the social function of technology and a truism that technology cannot be understood as an artefact per se, but should be seen in its economic, social and cultural context. I hope to have unravelled some consequences of this cliché by these cases of modernization strategies from above and from below. What made progressive Danish estate owners receptive to British agricultural implements and cultivation systems and what made peasants repudiate the very same technology transfer? As I have tried to show convertible husbandry and modern cast iron implements were also instruments of power and control which were used in the ongoing social struggle between agrocapitalism and tenant farmers. As illustrated by the two cases of opposing codes of technological education the whole question of technological power and control was embedded in opposing cultures. The case story about cream separators and the cooperative dairy movement was related to illustrate opposing views of technical and economic rationality and interests. The surviving tradition of cooperation in communities of small scale farmers established a modernization strategy from below by means of a technical artefact incorporating a cultural asset. I am convinced that further empirical studies of the reception of transferred technologies by various minor European cultures will make a substantial contribution to our understanding of the modernization process.

References

1. Chapter on Agriculture in *Teknik og kultur i Danmark, bd. 1 -1750-1850* (forthcoming)
2. Dan Ch. Christensen: *'Hvordan voksede planterne for 200 år siden?'* in *'Människa och miljö'* – xxi Nordiska Historikermötet, red. Lars Lundgren, Umeå 1991.
3. Dan Ch. Christensen: *'Agrar teknologi og social struktur. Historien om svingplovens indførelse i Danmark 1770-1850'*, RUC 1987 (stencil)
4. Dan Ch. Christensen: *'Tidlige Tærskeværker'*, in Bol og By – Landbohistorisk Tidsskrift 1991:2.
5. Morris Berman, *'Social Change and Scientific Organization. The Royal Institution 1799-1844'*, London 1978.
6. Dan Ch. Christensen, *'Oersted – A Romantic in Science and Technology'*, in 'Intellectuals reading Technology – Prodeedings from the Nordic Symposium 'Technology – Ideology – Culture', ed. Catharina Landström, stic no. 4, Gothenburg University 1991.

7. N.F.S. Grundtvig, *'To dialoger om højskolen'*, ed. Dan Ch. Christensen, ODIN 1983.
8. L.C. Nielsen is portrayed in *'Opfindernes liv'*, 2nd ed. Helge Holst, Cph. 1915
9. Claus Bjørn (ed.), *'Dansk Mejeribrug 1882-2000"*, Cph. 1982.
10. Erik van der Vleuten, *'Steam Engines and Technological Styles – Steam in the Netherlands and Denmark during the Industrial Revolution'*, 1992 (stenc. Eindhoven, Roskilde, joint Ph. D. project)

For additional information please contact the author through
TISC-project
Roskilde University
P.O.Box 260
DK-4000 Roskilde.

Ethics of Technology: The Irreplaceable

Peter Kemp

We talk nowadays about *a better life* for human beings. We are told that the important thing is to create more *quality of life*. But how is this better life, this quality of life, to be brought about?

A few years ago most people had no doubt that it was to be done through more and better science and technology. For the last couple of hundred years our culture has believed implicitly that science and technology are the key providers of quality of life. This faith was far from being pure illusion.

Indeed many older people in our society certainly know that material circumstances have created better opportunities for a happy life, and they often feel that young people do not sufficiently appreciate the material development that has taken place.

But the fact that it is no longer quite so obvious to many people that science and technology do create a better life is bound up with the problems created by the advances of science and the many new technologies, problems which science and technology cannot themselves solve, because these problems are not intrinsically scientific or technological. Such problems are not solved by greater scientific expertise, more technical experts etc. They make demands for quite a different form of insight, an ethical insight.

This is probably most apparent in the biotechnical area, where new medical and biological technology enables doctors to intervene in the most intimate areas of human life, right from conception and birth to sickness and death.

In general it may be said that although for a long time scientific and technological developments seemed to be sufficient for creating a better life and thus seemed to lead away from ethics, in our day they are leading us towards ethics. Today they have made ethics a practical necessity.

I. Ethics and action

Ethics deals with the good life and the good act. Therefore it is not possible to arrive at any clarification of the content of ethics without first determining what human action is. This question has been badly mishandled by many thinking people – scientists and even philosophers – who in modern times have tried to understand human action as a form of process on a par with physical or (bio)-chemical processes or as a form of function on a par with technical or mechanical functions. The ever-growing literature attempting to "explain" man as an especially complicated computer is a typical contemporary expression of the kind of thinking that reduces man to a machine.

But if ethics is to have any meaning and if the questions which science and technology pose without being able to answer them are not, in the final outcome, to be dismissed with scientific or technical answers which in the long term can only disappoint human beings, it will be necessary to clarify some of the most important features which separate a human action from a process that is scientifically observed and from a function that is technically performed.

Primarily, there are *motives* for an action, whereas we speak of replicable *causes* of a physical event. A motive is the reason for an action by which the agent allows his decision to be influenced, whereas a cause is a force or state already existing, produced, or spontaneously emerging, which brings about the process or event.

For instance, I have a sick friend in hospital. I am busy with my work, but I nevertheless decide to visit my friend. In this way I make his illness into a motive for the action. The illness is not a compulsion to act in a particular way, but I choose of my own free will to allow affection for my friend in hospital to outweigh the requirement to do a particular piece of work in the time I set aside to visit my friend. Thus a human being has a will to determine his actions, and this will is not compelled to act according to previously given orders, since it creates for itself the importance which it ascribes to what is happening: my friend's illness is given importance as a reason motivating my visit to the patient in hospital.

In contrast to motive, the cause which has an effect in the physical world is not a motive which man can create for his action. It is conceived as being the cause of the effect regardless of whether anyone produces it or not and regardless of whether anyone ascribes this importance to it or not. Also, the same cause is conceived as always leading to the same effect if the conditions for the process are identical. In this way the effect is *determined* by the cause, whereas an action is not determined by the motive, but *allows itself* to be determined by it. It must of necessity be "indeterminate" by everything outside itself (i.e. of all causes), since the agent, even under the very same circumstances, may choose to motivate his action differently: perhaps because he thinks he made the wrong choice before and now wishes to see what the result will be if he acts differently, even though the conditions are the same.

In addition, the importance of *circumstances,* like the motive, is not something given in advance. The agent certainly does not create them at all, insofar as they are physical and psychosomatic conditions which form chains of cause and effect, and usually indeed not very much if they are social conditions, but they may be conceived as being favorable or unfavorable. As an agent, a person at least decides the meaning or significance he wishes to ascribe to the compulsion to which he is subject and to what he has done because he was "driven to it" by instincts, dispositions or reflexes (e.g. of fear), and often a person can choose the circumstances on which he wishes to base his action and those from which he wishes to distance himself or which he wants to try to change. So we see some circumstances as being favorable, or we choose to oppose the conditions we receive as given, in order to alter them, and then we regard the given circumstances as unfavorable to the action. In that sense a person is "master of the situation".

The foregoing already makes it apparent that action presupposes an agent, who can carry out a work which is *the person's own work*, so that he or she may be held responsible for certain consequences of the action. Being responsible or answerable means that, in reply to the question who carried out the action, one can *answer* "I did" or "It was I".

In other words, action presupposes a human *person*. Although the word "person" originally meant a mask or a rôle, in modern usage it has come to denote that which is behind the mask, the living being himself, who is self-aware as one who acts independently and who has an intention with his acting.

Not any and every living being is a person, for what is alive – e.g. a plant or animal – is not necessarily self-aware as independent or *autonomous.*

This autonomy, which is cited by Kant, for instance, as a basic precondition for all ethics, is not established by a living being until it becomes capable of distinguishing itself from others or differentiating itself from everything and everyone else.

This still applies even when the person's speech and action are determined in a psychoanalytical sense by the Unconscious (Freud) or the Other (Lacan), so that the Ego is not master of its own house. Psychoanalysis rightly denies that humans are born autonomous, but it aims at a liberation of the subject from the Ego's illusions and narcissism and thus at as much automony for the subject as possible to speak, act, and love. The subject (we say: the person) – through insight into his own desire – can liberate himself from his blighted ego.

It may therefore be said, as Hegel did, that the personal element (which he called "the Spirit" or "the spiritual") is "the negative". For it consists of the ability to *say no* or opt out. But nobody can say no to one thing without saying yes to something else (even if it is only suicide to which one says yes) which means that the personal is also something positive, an idea or conception of something one wishes to actualize or about things which one finds valuable in the existing world. It is initiative.

It follows from this that an action that is carried out by a human person is neither a sheer process determined in a cause-effect relationship, nor a technical function, which of course has a human instigator (the technician and/or planner) but which from then on, if all preconditions and circumstances are known, carries on the predetermined activity. The agent is neither a (physical or biological) thing or a machine, nor a robot which performs the movement it has been given or is ordered to do, but a being who is aware of himself as a will to renounce one thing and at the same time take responsibility for something else.

A human person is also a being who – as Hegel pointed out – has a need for other people's "recognition" of the contribution he or she makes to the wellbeing of society or the community as a whole. Therefore the actions of human beings are done with a constant eye to the recognition they can expect from others and the kind of person they wish to be recognized by. The genuine person is not a little ego shut in upon himself, but a Self who opens out towards someone else, other people.

All these determinations of the agent exclude the possibility of understanding a human being as a machine or robot. It is of course true that a machine – for instance the computer – may be used as a model to help us understand certain features of human beings as biological and thinking beings.

This is hardly surprising as all machines, robots, or technical systems have been created by humans to replace and extend some of the movements of the body or the calculating functions and text manipulations of the brain, which human actions would otherwise have to do for themselves. So machines always resemble humans on some point.

But if anyone deduces from this that there is basically no difference between man and a machine, and tries to support this assertion by pointing to the similarity between human thinking and the calculations and word-processing of the computer, this is not only a wrong conception of human beings. It is also an attack on ethics.

We have here reached the heart of the matter: *the inevitable connection between our conception of humanity and our moral practice.*

The answer to the question whether man is a kind of machine or not can never be a neutral description. For instance, it is an ethical decision whether "computers are alive" (Geoff Simons), or whether human beings, in contrast to computers, may be accounted autonomously acting volitional beings who are responsible for what they do and who open themselves to other people's recognition of their autonomy. Our conception of what human beings are has ethical consequences as well, for we naturally treat human beings as we conceive them to be: if they are conceived as being machines, there is nothing to prevent us with "a good conscience" from treating them as machines, i.e. relating to them cynically as mere means to the end to which we ourselves aspire.

In doing this, however, we violate the idea of the individual's *irreplaceability*, which is the basic idea of modern ethics.

The consequence is that no philosophical ethics can be a neutral description, but the view of man and society implied by the philosophical text is in itself an ethical stance for which the author is responsible, as for any other ethical act.

II. Ethics of technology

An *ethics of technology* unites the two concepts of ethics and technology. What can these two things have to do with each other?

The particular feature that separates modern society from that of the last century is the technological transformation of everyday life that has taken place. It has become apparent that modern technologies like nuclear power, the biotechnologies and computer technologies pose a number of questions about how we can live with them in everyday life, questions to which the experts within these technologies themselves have no answers and which maybe they neither can nor should solve, for they are not technical questions at all, but questions about human attitudes to technology and the choice of the right technology, of the right technological development.

Examples of these questions are: Should we run the risk of another Chernobyl by maintaining or even expanding nuclear power? Should we treat animals purely as material for food production? Should we aim to reproduce as human beings by replacing sexuality as the precondition for producing children with the test tube and replacing the mother's pregnancy with the artificial womb? Should we set up electronic monitoring of all workplaces, in shops and in streets, not to mention homes? All these questions are ethical questions. They are not opposed to technology as such but they are concerned with the right way of living with other human beings in relation to the new technologies.

No doubt we have discovered these questions only because the technological optimism which we inherited from the nineteenth century has been dealt some shrewd blows, which have caused it to collapse.

III. The collapse of technological optimism

The nineteenth century was optimistic. In those days people believed in progress, i.e. that the development of modern technology and science would necessarily and automatically lead to a better society, to a better life for all human beings. Both Marx and his opponents, both socialists and liberalists, agreed on this faith in progress: Of course we would have to endure a great deal of suffering, but finally we would attain the "realm of liberty", thanks to the enormous technological development. Karl Marx, in the French version of *Capital* which he himself approved, expressed it thus: "In society as in nature, rottenness is the laboratory for life."[1]

This meant that ethics, or the question of the good life, as the men of the nineteenth century saw it, would be solved by science and technology. Neither science nor technology was regarded as neutral; they were unquestioned goods or the good for mankind, and any criticism of science and technology was evil, reactionary, backward-looking. Criticism was attributed to a few religious people who maintained an outdated world picture, or to some anarchists who refused to bow the knee to God, the State or Science.

It has often been said that optimism was given its first mortal blow by the First World War. But it was cultural optimism, the belief in the power of high ideals over human beings, which suffered defeat then. Technological optimism continued. It is true that at the end of Volume 2 (1922) of his great work on *The Decline of the West*, which everybody was reading after the First World War, Oswald Spengler wrote that as machines become ever more human, ever more immaterial, ever less noisy (*immer geistiger, immer verschwiegener*) so they progressively deprive their makers of power, turning human beings into slaves of their own creation: "The wheels, rollers and levers are vocal no more. All that matters withdraws itself into the interior. Man has felt the machine to be devilish, and rightly. It signifies in the eyes of the believer the deposition of God."[2]

But in Spengler's view machines were nevertheless on the side of life against the power of money, high finance or capital. Machines were on the side of socialism.

It was not the world wars as such that shook people's confidence in technological advance. Classic European culture had proved its shallowness: beneath the surface man was not so good as he should be. But technology, which man himself had created, was a material force people could believe in, even when they could no longer believe in the spiritual power of human beings.

Technological optimism collapsed as a result of a number of shocks which people have been receiving from technological development itself since the 1940s.

First Shock: Hiroshima and Nagasaki

The first shock was the dropping of atomic bombs on Hiroshima and Nagasaki in Japan. The aspect which shocked the person of European culture (particularly western Europeans and Americans) was not really that the bombs were dropped on somebody (for everyone breathed a sigh of relief that the war was thus brought to an end), but that the best brains of society, who presumably were also its best men, had been able to produce a weapon that could exterminate all life on earth.

The making of the atomic bomb began to call in question the dogma that science and technology, when left to distinguished scientists and technologists, would automatically create a better society. It was in order to uphold this dogma, to maintain this image, that Niels Bohr was prompted by his best conviction to pay visits to the leading statesmen of the world with the aim of bringing about an international agreement on the control of atomic weapons.

However, something more effective – at first – was the use of nuclear power for what was called peaceful civil energy production. This seemed to mean that the dogma could still be maintained. So in the fifties and sixties we saw a technological development carried along by a more uninhibited technological optimism than ever, with the creation of the highly industrialized welfare societies of the USA, Europe and Japan.

Second shock: Limits to growth

But then came the second shock: the ecological shock which was triggered off in 1972 by the report of the Club of Rome: The Limits to Growth. This asserted, with sober figures to back it up, that economic, technological and demographic growth could not continue exponentially without ending in collapse. There would be too many people, pollution would suffocate us all, and the world economy would get completely out of control.

The Limits to Growth declared that many of today's problems have no technical solution, i.e. "one that requires a change only in the techniques of the natural sciences, demanding little or nothing in the way of change in human values or ideas of morality."[3] Examples of problems without technical solutions are given in the report as the arms race, racial problems and unemployment, but also society's technological progress as a whole, whose exponential development must end in disaster if it continues unchanged.

The year after The Limits to Growth came the oil crisis, which made people realize that not even their energy supply was secure. Along with the conviction that growth could not continue unchecked, this crisis led to people beginning to talk about the necessity of controlling technology politically.

The initial outcome was that politically there was a greater concentration on nuclear power. This led to the great debate from the mid 1970s about the safety of nuclear power. In a number of countries, small groups of concerned scientists and technologists spoke up to demand more detailed studies of the consequences of expanding the nuclear power industry.

These critics found support from a number of university scholars in the humanities, from artists and writers. In this way, analyses were published of both the technical problems and the humanistic questions posed by the expansion of nuclear power.

Criticism focused on the following three main points:

Firstly, faith in mathematics. The critics attacked the calculations of probability which were intended to convince populations that the worst conceivable accidents would be so infinitely improbable and therefore in practise no problem. The report known as the *Rasmussen Report* became famous for this. That was before the accident on Three Mile Island in the USA in 1979.

Secondly, faith in "the technical fix". The critics attacked the naïve confidence that an acceptable technical solution to the "waste problem" in particular was within reach. At that time, over ten years ago, people were saying in Denmark, for instance, that they would need at least ten years to find an acceptable solution. Today Denmark has completely given up any idea of finding one.

Thirdly, faith in the expert. The critics attacked the arrogance of the experts, both their scornful treatment of all lay people who voiced an opinion in the public debate, and also the way they branded critical scientists as untrustworthy, unqualifed traitors to collegiality and the scientific community. On the other hand, these experts willingly delivered their opinions with the weight of scientific authority in areas where they were not experts, for instance a Danish reactor physicist opposed the opinion of geological experts about the geological possibility of storing radioactive waste underground in Denmark. Such arrogance has become harder to practise today.

Third shock: The big disasters

All in all, this debate sparked off general scepticism of all experts, not just in the nuclear power field but throughout society. This was bound up with the third shock: the visible wave of major accidents in industry, not just in the nuclear power industry, but also in the chemical industry, ranging from Seveso in 1976, Three Mile Island in 1979, to Bhopal in 1984 and Chernobyl in 1986. The accident on Three Mile Island is now known[4] to have been far more threatening than people were led to believe in 1979, and some years after the Chernobyl disaster experts have discovered that the radioactive cleanup of the area has been a failure;

instead they have demanded a fresh evacuation of over 100,000 people.[5] Not only the disasters themselves, but the exposure of the many attempts each time to conceal what had really happened, have seriously rocked the pedestal on which modern scientific and technological researchers had hitherto been placed.

In addition, the actual theoretical debate about nuclear power marked the breakthrough of the view that technological development is not something given, inescapable, but that it involves choice and that a choice of technology is a choice of society. The choice of technological and scientific development is political, and insofar as political life entails an attitude to what the good life is, the choice is also ethical. It is a choice of what life we ourselves will put up with, and what life-opportunities we wish to give to posterity.

Fourth shock: Test-tube babies

The fourth shock was brought by "test-tube babies" and all the possibilities which new medical technologies open up for intervening in the human reproductive system. The manipulation of genes and freeze techniques thus pose questions as to how far it is defensible to go in charting and controlling human biological life. A French researcher, Jacques Testart, who was one of the two scientific "fathers" of Amandine, the first French test-tube baby, thus wrote a book on *The Transparent Egg* (1986), in which he declared that he would now refrain from some forms of experiment which seemed to have no prospect of serving any respectable human purpose

The whole development of medical technology has abundantly confirmed the vision of the future in Aldous Huxley's novel Brave New World, which depicts how it would be possible to overcome all human sufferings by using medicinal substances, especially what the novel calls soma.

Fifth shock: The computerization of society

The fifth and most recent shock has been brought about by the informatization of society, which does of course bring us the advantages of automation, but also means that we have less and less control over more and more areas of our own lives. Computerization promotes bureaucratization, the turning of human problems into impersonal cases decided by inscrutable authorities. Thus the taxation system has become an impenetrable jungle, which often can not be fully understood by the local tax officials whose job is to give guidance to taxpayers who have tax problems.

In addition, computerization promotes opportunities for the monitoring of people's behavior patterns. In this way we approach the kind of society which

George Orwell described in his novel *1984*, in which everything is controlled by Big Brother, who is always watching you!

*

So technological optimism has been shaken by a wave of shocks which may tempt people to go to the opposite extreme and become anti-technical. Then two wings arise in society, and a Danish expert in reactor technology did feel he could deliver a telling blow against opponents of nuclear power by talking about an up-wing and a down-wing: the up-wing comprised supporters of nuclear power, the down-wing was all the others.

On a higher plane, the French philosopher Jacques Ellul has expressed a pessimistic attitude to technological development in his book "The Technological Society", which appeared in French back in 1954 ("La Technique ou l'enjeu du siècle"), but did not make much of an impact until the first American edition appeared in 1964 just when Herbert Marcuse in the USA was beginning to criticize the "affluent society".

In Ellul's view, technological development is a destiny we must submit to, but against which we can do little. Instead, we must concentrate on the inner life.

But before we take refuge in the arms of technological pessimism, it might be worth considering whether man does not have opportunity to bring about a more humane development of the technologies.

The precondition is, of course, that it is possible to view the technologies from other angles than purely technical ones. This means that the question is: Is man in his essence something other than a technician? Can we control and direct our technical life with non-technical yardsticks?

IV. Technical man

Let us first look at what a human being is as a "technical being", and then look at whether a human being is also something other and more than that.

1. Technics as a form of life

Martin Heidegger (1889-1976), for instance in the writing he published in 1953 "Die Frage nach der Technik" ("The Question concerning Technology") and in his book on *Nietzsche* (1961)[6], depicted the sense in which man, that is any human being, is a technician. Man is a technician, he says, when he seeks to find a firm footing in existence, to adapt, and therefore seeks to take possession of things. Then he knows something about what he has to do to adapt himself to things. This knowledge is his technical knowledge. In this way technics is not primarily a question of being able to produce and manufacture, but of being able in everyday life to comprehend things, use them, and live with them.

Technics is a physical knowledge about what we have "ready to hand" (zuhanden) or can regard as being "present at hand" (vorhanden), and is thus knowledge about the relationship between man and things. Technics is therefore not only a tool for an action. Technics is an insight into how man may exist with things. Technics is a way of being, a form of life, in which man has understanding of something, is capable of something, can do something.

Heidegger sees technics not only as mastery of things, but a disclosure of the truth about man's life with them, i.e. the truth about human existence as care *(Sorge)* for oneself, for one's life with things and with other people in conditions of finitude.

2. Technics as the production of a work

It is not until this basis exists that technics (the technics man uses to live with things) becomes a production of works, either craftsmanship or works of art. In Greek the word "technics" originally meant not only that which we nowadays understand as technics, something functional which makes our action more effective. In Greek technics also meant art. Or rather, there is no essential difference to the Greek mind between producing a work of craftsmanship and producing a work of art, but a human being's work may be a work of greater or less success. If it is a great work, then it was made by a great artist.

The work, and of course at best the good work of art in sculpture, painting and poetry (in Greek means producing or creating, whereas poem, means the product, the thing created) expresses the truth (which technics contains) about what it means to exist with things. According to Heidegger, what flows from this is the importance of art and especially of poetry for human beings' understanding of their own lives. That is why he often refers to Hölderlin's line: "Poetically dwells man upon this earth."[7]

3. Technics as a tool

The works which man produces can be used as tools for producing other works. This means that technics becomes that which we nowadays often envisage when we speak of using technology, of introducing new technology. In its modern form, says Heidegger, this technics is a storing, or "enframing", of energy. Thus, he asserts, it amounts to a challenging of nature, whose forces we are seeking to master ("regulate and secure"). As an existent being, man is without substance, solidity, consistency. It is for this very reason that he seeks to save himself by becoming, through "regulating and securing"[8], the lord of nature, of things. He attempts to find permanency as "lord of the world". Technics is the tool for this.

4. Technics as the basis of natural science

Now Heidegger asserts that modern science has arisen on this basis, i.e. on the basis of the existence of technics as a form of life, as a work and tool. The contrary is not the case: science has not laid the foundation for technics. Science has been made possible because man is a technician, forming things, creating works and using them as tools. Through natural science he seeks to improve his potential for subduing nature.

Heidegger finds that this view is supported by the mutual relationship between the theoretical results of science and its instruments. It is by applying technics, i.e. tools and experiments, that man attains scientific knowledge, and the nature of the instruments conditions the knowledge that can be attained. Conversely, new knowledge opens up the possibility of producing new instruments or using old ones better. Thus Heidegger has propounded the view that the history of technology is not a mere component of the history of science, but rather vice versa.

It might perhaps be better to say that technology and science each have their history, which originally had little mutual contact, as long as technics was the skill of the craftsman and artist, his savoir-faire, his know-how, while natural science was restricted to speculation about the physical world and mathematical cogitations that were rarely translated into technics.

Gradually however the two forms of history have encroached more and more upon each other. Technology has been scientized, its development being conditioned by scientific research. Conversely, knowledge of the physical world has been increasingly technologized, as it has become dependent upon the instruments, measuring apparatus, machines and systems being developed.

5. The concept of "technology"

Some idea of the import of this technologization may be given by a brief etymological analysis of the word "technology" itself, which is made up of two Greek words, *techne* and *logos*. Logos originally meant speech, lore, knowledge, understanding, insight, whereas as we have seen techne originally meant skill or proficiency in arts and crafts. Therefore the original meaning of the concept of "technology" was knowledge about skilled craftsmanship. But such knowledge is in itself a skill, a theoretical skill. So the concept has developed to mean theoretical skill concerning craftsmanship.

Through the application of scientific insight, this theoretical skill has expanded opportunities for technical proficiency (technique or technics), so that nowadays it covers not only human production and processing of things with simple tools, but also the use of machines and science-based systems. In this way the word "technology" has acquired its modern meaning: it means not just technics, but the application of a combination of science and technics.

So when, in modern society, technology is an element in an industrial or organizational project, we often see the word used about the project itself and the social structures and relationships which it implies. A technology in this sense may be the use of genetic engineering to develop new types of grain, the use of nuclear power to produce energy, the application of computers to office work, etc.

6. Critique of Heidegger

Thus a distinction must be made between technics and technology, which Heidegger does not do. He uses the German word "Technik", and this is indeed translated by "technology" in the English version of his works. In this book we often use "technics" instead of "technology" as the most appropriate way of rendering Heidegger's thinking.

The reason that Heidegger does not distinguish between technics and technology is probably that he is interested only in self-reflection upon man's technical activity as such, not in the problem of direction and control posed by modern technology, for increasingly it is no longer the individual's work or tool, but a collective phenomenon over which human beings risk losing control.

The difference between classical and modern technics is seen by Heidegger solely in the fact that in modern times technics has forgotten to be an art and has become aggressive; it has become challenge, making demands upon nature for energy which can be extracted and stored ("enframed"). In other words, this technics has become a manipulation of nature. But that is a deficient view of what modern technology is. Heidegger overlooks not only that technology is also – or may be – a game in which we take part. But by his definition, all computer technology and in general all social technologies (organizational technologies, Taylorism, management technology etc.) fall outside of what is technology.

But modern technology is not just the technology of tools and energy by which man relates to nature, but it is very much information technology and social technology, whereby human beings communicate with each other and relate to each other.

It has been said (by Lévinas for instance) that in this way Heidegger is a "peasant philosopher". This has prevented him from understanding *Mitsein*, the relationship between other human beings. For him therefore, technology has been reduced to a relationship between existence and nature (the earth, die Erde). And modern technology is then an "enframing" (Gestell) of that nature whose forces or energy "are put in place". This is what the "frenziedness of technology" consists of.[9] We must – he thinks – leave this technics to be what it is. To save ourselves from it, is not a question of directing or controlling it better than before, but of opening ourselves up to the essence of technics, creating technics in the form of art, and thus more clearly recognizing ourselves as existent.

V. The sensitive human being

Although in his main work, *Being and Time*, 1927, § 29, Heidegger wrote a detailed account of state-of-mind ("Befindlichkeit") and attunement or mood, this philosopher of existence, oddly enough, knows only technical man. Therefore ethics never finds any place in his philosophy. For ethics presupposes – at least if it is to impinge upon relations with other people – something other than the technician. It presupposes the sensitive human being.

If ethics is a necessity and is not to be reduced merely to a special form of technics, we must assume that man as a physical being has both a technical and a sensitive existence: that he produces, adapts, and shapes things, and also senses, perceives, grasps and interprets the world in surrendering to it and receiving from it.

This sensitivity is not inactivity, but an activity which is not firstly directed outward to the world around, to things, but is directed toward ourselves as percipient and acting persons, and thus to our own situation. It is the activity whereby we shape ourselves and form for ourselves an attitude, so that thereafter we can shape the world on the basis of this attitude.

However, sensitivity may itself have two directions. It may either apply to our own situation, or it can apply to our relations with other human beings.

1. Being sensitive

In the former case it need not have anything to do with ethics. It is concerned with how man not only shapes the world, but also how he receives it with his senses and opens himself up to it with his needs (a feeling of lacking something) and feels that he is present as a body.

The experience of sense-perception gives the basic feeling of pleasure at existing, joy of living.

Need gives the basic feeling of being attracted by something, having an appetite for something, while there is something else by which one is repelled or against which one has an aversion.

And bodily presence gives the feeling of being "flesh", to which may befall either pleasure or pain.

When need is reflected as the person's will to reach a goal in a world with many obstacles, it is felt as passion, which may be positive longing or negative dislike.

I have given a more detailed account of the whole of this feeling aspect of man in my treatise *Théorie de l'engagement*, 1973.[10] The important thing here is to make it clear that man can not be understood merely as a technician, not even as the existential technician of Heidegger's account.

2. Feeling for someone

The only reason that human beings can feel for others is that they are able to feel that they are bodily alive in the world. Ethics presupposes the feeling of being present.

Ethics itself is a vision of the good life. This requires not only that one is sensitive, but that one has feelings for other people. In our culture, at least, it would be hard to imagine a good life without some kind of community with others.

Aristotle (384-322 B.C.), who was the first to work out a doctrine of the good life, i.e. an ethics, conceived this ethics to be part of political life and thus a social phenomenon. His *Nicomachean Ethics* is a kind of extended introduction to his work on *Politics*. Thus ethics has to do with the right relation to other human beings: with our actions toward them, with our reactions to their actions, with our cooperation with them. Ethics is therefore not just a question of friendships, but a question of societal organization, of government, of politics.

In Judaism too, before the emergence of Christianity, ethics was a question of the formation of society. The Ten Commandments were originally rules for social living. At that time there was no sharp distinction between ethics and law.

But with Christianity a radicalization occurred (which might be called an "impassioning") of ethics. This came about when Jesus introduced the idea of the absolute value of the individual human. Some words of Jesus indicate this. For instance, the words about the lost sheep: "If a man has a hundred sheep, and one of them has gone astray, does he not leave the ninety-nine on the hills and go in search of the one that went astray? And if he finds it, truly, I say to you, he rejoices over it more than over the ninety-nine that never went astray" (Matthew 18:12-13 RSV). Also the famous parable of the prodigal son who returns home and is fated, "for this my son was dead," says the father, "and is alive again; he was lost, and is found" (Luke 15:24 RSV). Finally the well-known utterance, "For what does it profit a man, to gain the whole world and forfeit his life?" (Mark 8:36 RSV).

This makes ethics something much more radical than law, which orders the external relationships between people.

Today we may express this radical thought as the idea of the individual human being's irreplaceability. This idea was first seriously propounded philosophically, detached from ecclesiastical language, when in 1785 Immanuel Kant in his book on the Grundlegung zur Metaphysik der Sitten (*Groundwork of the Metaphysic of Morals*) maintained that the main idea in ethics (practical philosophy) was that a human being must act in such a way that at any time he respects the human person as an end in himself and never merely as a means.[11] Of course, we all treat each other as means to our ends, as we derive benefit from each other in order to achieve what we want. But if we treat others purely and simply as means to our ends, they have been reduced to nothing but material for our actions. Then we are not regarding them as ends in themselves, as being irreplaceable.

It is interesting to note that Kant was saying this during the period when industrialization was beginning to change the pattern of society, so that people had to leave their trades and find work in large factories, where they were treated as functions that could be replaced. Kant's practical philosophy is thus a reminder that the value of a human being is not an economic or technological value (what we call a price), but an immeasurable value: the irreplaceability which manifests itself in that the human being we lose can not be replaced by another without loss, that he is indeed not measurable. We feel for him in the way that the loss of him gives a feeling of emptiness, an emptiness that it will never be possible to fill.

Notwithstanding that a person in a factory, an office, or business, can be substituted by another worker or staff member taking his place, a human being can not, according to the radicalization of ethics that has taken place in our culture, be replaced. It runs counter to our feeling for the other person to regard him merely as a means.

A Canadian philosopher, Christopher Hodgkinson, has expressed this in these words, "No one is indispensable. Everyone is irreplaceable."[12]

Thus the modern ethical vision is that every human being is unique. First off we might think that this could be understood biologically (in the sense that each has his own DNA formula). But this is inadequate. In the final analysis the uniqueness is something psychological, or better, existential. Otherwise human beings with the same genetic code (e.g. identical twins) would not be unique, but each of them is – ethically-existentially – irreplaceable. For they are not merely mutually interchangeable specimens. If one of them dies, the loss is irreparable. It is of no help that another being exists with the same DNA code.

VI. Human dignity

The idea that lies behind our speaking so much nowadays of respect for human dignity and requiring respect for human rights is indeed this very idea of irreplaceability.

This idea of human dignity has a long history of development. It goes back to an old idea in our culture, that man can raise himself up above nature. The Greek philosophers spoke of thinking, which raises itself above purely vegetative and sensual life, and the Jews envisaged that at the creation man was set to have dominion over the other beings in nature. Here man has his dignity – which to the Jews, though, was subordinate to God's majesty – by virtue of his dominion for good or ill over nature.

This idea was taken over by Christianity and was radicalized by the philosophers of the Renaissance, particularly Pico della Mirandola, who in 1488 published his passionate speech "De hominis dignitate" (*On Human Dignity*).[13] In this he said that man surpasses everything else in the world. For he can break with the limitations that apply to all other living beings, who are able only to develop into that which nature has predetermined. By his will man can himself set limits to his life. He is his own "sculptor" and he is the architect of his own world. This means that he can allow himself to sink down to beastly forms, but also that he has power to raise himself up to a divine life.

In this way the idea of man's dignity has become the idea of his sovereignty by virtue of his will and thought. This finds expression in Kant's idea that man's independence – his autonomy – is the ground of "the dignity of human nature and of every rational nature".[14] This autonomy consists in "his capacity to make universal law, although only on condition of being himself also subject to the law he makes,"[15] i.e. in the ability to ordain and set limits for himself to the way in which he is to live his life.

This marks the difference between dignity and a price. If something has a price, says Kant, "something else can be put in its place as being equivalent; if it is exalted above all price and so admits of no equivalent, then it has a dignity."[16] In this way human dignity involves that any human being is irreplaceable.

In other words, the difference between price and dignity is that price is a relative value, whereas dignity is an "intrinsic value". The latter is not a means to a higher end, but is an end in itself.

It is in this context that Kant propounds the idea that we must never treat a "rational being" as purely a means, but always treat it as an end in itself, and thus always "as a supreme condition restricting the use of every means."[17] With this he combines Christian ethics (about paying heed to every human being) with the idea of human dignity by virtue of being able to raise itself by will and reason above nature.

VII. Ethics and norms

The reason that we often have difficulty in formulating the importance of ethics for the choice, planning and development of technology and science or for management and cooperation in business is probably that ethics is thought of as a given set of norms, of prohibitions and injunctions about what we should and should not do. If this is so, it is hard to handle, for particularly in new situations created by new technologies we often find that the rules hitherto applicable no longer give us any guidance.

But in itself ethics is not norms, not injunctions, prohibitions, in other words not moral rules. It is a view of humanity, a vision of what the good life is, as Aristotle expressed it. A vision of how we should live. A view of humanity which, to some extent, has already found expression in practical attitudes and in all our narratives about them, but which nevertheless remains a task and has to be continually re-expressed. Both reality and ideal.

A vision like this gives guidance when we are faced with a technological problem with human aspects, but it can never dictate a solution; it takes into account that new situations can arise in which the ready-made solutions are no longer adequate.

When we make norms into the basis for decisions, we are making what is normal, i.e. customary, already approved, into the basis. This has been strongly emphasized by the French philosopher Georges Canguilhem in his book *Le normal et le pathologique*.[18] He has shown that the concept of the normal is a "polemical concept" with which to criticize others, and that norms aim to "normalize", i.e. demand that differences be ironed out and that deviancies be removed.

Norms may be unexceptionable as long as we do not make them absolute and immutable. Think of the norm that a married couple should have children; this norm is alright insofar as it ensures the continual renewal of mankind, but if it is made absolute, then all means are permissible for producing a baby, regardless of the psychological and physical cost to the parents (or the mother, at least). Or think of the norm that medical science should prolong life: if this is made absolute, we find ourselves in many cases with bodies which live without souls, or which have become specimens with which the soul can not endure to live.

Or think of the norm that we should extract energy from nature in order to improve society for everyone. If this is made absolute, we become willing to run the greatest risk, e.g. by expanding nuclear power stations, regardless that it may end in disasters like the one in Chernobyl.

Or the norm that we should ease the life of other people by automation. If this is made absolute, the ordinary citizen is, as we have mentioned, deprived of control over his own life, as is happening in the economic sphere, where more and more things are being settled through electronic data processing, when the computers work everything out for us and only a few specialists know what is really going on.

Norms can become outdated. It must be possible to reshape them or reject them. Then of course we can set new norms, but where do we find the basis for doing this? On what basis are we to act in situations where the old norms are inapplicable and new norms have not yet been established?

We need something other than norms to build upon in non-normal or abnormal situations. We need a view of humanity that can be actualized in an attitude. We need an ethic.

*

Since this view of humanity is more basic than the norms, we must thus make a distinction between ethics and morality. By ethics we may understand the actual basic conception of what is the right way to live, whereas by morals we may understand those norms or moral rules into which we attempt to translate ethics in order to maintain a certain amount of good order in our existence.

This fits in very well with the fact that the word "ethics" comes from the Greek *ethos*, which originally meant "dwelling", "home", and the like and was later used by Aristotle about the character that is nurtured in daily living, while the word "moral" comes from the Latin mos (plural mores) and was used by the Romans (Cicero, for instance), who desired order in family and society. Both words came to mean the same thing ("custom" "usage"), but they have developed in different ways, so that ethics is applied mostly to fundamental philosophical reflections upon the good life (in which man feels at home), while morality is applied mostly to the limits we set for each other's behavior (by desire and will).

Thus morality is the same as the normal, the conventional, slightly boring but also problematical, while ethics may lead to the extraordinary, the unique, and help us in new and unforeseen situations.

VIII. Universalism and pluralism

Now many people assert that today there can not be any generally obligating ethic. For has not each person his own ethics? However, the question is what we understand by ethics and whether it is not being confused with morality.

There is reason to assume that plurality of morals (the fact that there are many kinds of morality depending upon traditions and areas of living) does not exclude the universality of ethics, i.e. the universal applicability of an ethical vision. Although many norms crumble, there may well be a modern ethos, i.e. an ethic which can be traced as the basis of all important international declarations in our day about human rights and respect for human beings.

We shall do no more here than refer to the last passage of the World Medical Association's Declaration (1975) called the *Second Helsinki Declaration*, abbreviated to *Helsinki II* : "In research on man, the interest of science and society should never take precedence over considerations related to the wellbeing of the subject" (see Chapter 6 of this book). On the basis of this ethic, this vision of the individual person's value, the Declaration requires "informed consent" from people it is wished to use in medical experiments.

Conclusion

With modern technology human beings have acquired new opportunities for treating others purely as means. The technological manipulations have in themselves no built-in limits. So our idea of this limit has to be obtained from elsewhere. That is why ethics is neither a science nor a technology but the vision of experience and narrative imagination about what man and his world is. Ethics is a view of man, indissolubly bound up with a view of society and of nature.

Hence the maxim: THE WAY WE REGARD A HUMAN BEING IS THE WAY WE TREAT HIM – AND VICE VERSA.

The boundary line of manipulation runs where our view of man sets it. Thus ethics is inconceivable without an ontology, without a conception of "being", i.e. of what is. It is an imperative ontology. Conversely, no ontology is innocent, however much it may attempt to place itself on a meta-level, as Heidegger did in Being and Time and as Sartre did in *Being and Nothing*. Even when an ontology gives no direct answer to what one ought to do or refrain from doing, it can not speak about human existence or human behavior without hinting whether or not man is on the same level as manipulable objects, i.e. whether he is "on hand" (Heidegger) or whether he is a conscious project (Sartre). If the ontology entails that man is not an available object, then the legitimization of those actions whereby people manipulate him, treat him as purely a means, becomes invalid. And then the ontology at least means an openness to ethics.

Yet any ontology which tries to remain on a purely pre-ethical meta-level is not without danger for ethics, because it attempts to give the impression that ethics is something secondary to the pure thought of the philosopher and that practise is subordinate to theory. So openness to ethics may be transformed into openness to pride, ruthlessness, oppression. In this way the Nazis and Heidegger himself in the 1930s were able to understand existential philosophy as open to the idea of the victory of the strong over the weak in a life-and-death struggle for technological supremacy and to the excellence of the "dangerous" action regardless of the sacrifices it might give rise to.

Technology is a tool for human beings' shaping of things and each other. It therefore is separate from man himself, who will not be reduced to a mere tool of anyone else, but seeks a community which can stretch far back into the past (maintaining the memory of the exploits and sufferings of the departed) and far ahead into the future (in concern for coming generations).

But if, notwithstanding, we adopt a reductionist view of man and regard others as mere tools for reaching our own goals, it has no meaning to deny that they may be used and treated like such tools. In other words: Only if man is not himself technology, can we speak of a limit to treating him in the same way as technology. Only if man is not a machine does ethics have a meaning. Only then is an ethics of technology meaningful.

And conversely, ethics is meaningless if man is a machine. Ethics requires that the individual human being is regarded as the irreplaceable, and that human community is of a different kind than a technical system, in other words that giving and receiving are something other than function and manipulation.

The ethics of technology is not hostility to technology. On the contrary. From first to last it must be a defense of good technology, which does not reduce the Other to a tool, but sets limits which both give promises about new opportunities for a richer existence and also forbid the murder of our life with others.

This paper corresponds to some sections from Det Uerstattelige, Spektrum, *Copenhagen, 1991. Translated by David Stoner.*

Notes

1. Karl Marx: *Das Kapital;* French translation (corrected and approved by Marx): *Le Capital,* Ed. Sociales, Paris. I.2 (la Machinerie), p.168.
2. Oswald Spengler: *Der Untergang des Abendlandes,* I-II (1918-23), Munich, Beck, Vol.II. p.625. ET: *The Decline of the West,* trans. C.F.Atkinson, George Allen & Unwin, London, 1932, Vol.II, p.504.
3. Dennis L.Meadows et al. (The Club of Rome): *The Limits to Growth,* Universe Books, New York, 1972. p.150.
4. *Le Monde,* Nov. 2, 1988.
5. *Le Monde,* Aug. 9, 1989; *Der Spiegel* Apr.23, 1990, *Frankfurter Allgemeine Zeitung* June 8, 1990.
6. Martin Heidegger: *Nietzsche* I-II, Neske, Pfullingen, 1961. p.96-98. ET: trans. M.Krell and D.Farrell, Harper & Row, New York, 1984.
7. Martin Heidegger: *Die Frage nach der Technik* (1953) in *Vorträge und Aufsätze,* Neske, 1954, p.39. ET: "The Question concerning Technology" in Basic Writings, Harper and Row, New York, 1977, p.316.
8. *Ibid.,* p.31; ET p.309.
9. *Ibid.,* p.39; ET p.316.
10. Peter Kemp: *Théorie de l'engagement* I-II, Editions du Seuil, Paris, 1973, I 26-27.
11. Immanuel Kant: *Grundlegung zur Metaphysik der Sitten* (1789), Felix Meiner, Hamburg, 1952, p.54. *Kant's gesammelte Schriften,* Akademie Ausgabe Band IV, Berlin, 1903., p.429. English translation in H.J.Paton: *Groundwork of the Metaphysic of Morals,* Harper and Row, New York, 1964. p.96.
12. Christopher Hodgkinson: *The Philosophy of Leadership,* Basil Blackwell, 1983, p.111 & 229.
13. Giovanni Pico della Mirandola: *De hominis dignitate.* ET: *On the Dignity of Man,* trans. C.G.Wallis, Bobbs-Merrill Co., Indianapolis, 1965.
14. *Grundlegung,* Ak. Ausg. p.436. ET: p.103
15. *Ibid.,* p.440. ET: p.107.
16. *Ibid.,* p.434 . ET: p.102.
17. *Ibid.,* p.438. ET: p.105.
18. Georges Canguilhem: *Le normal et le pathologique,* P.U.F., Paris, 1966, p.175ff.

Comparative Studies in European History of Technology

Kristine Bruland

The purpose of this paper is to consider the potential contribution which comparative studies in the history of technology can make to our understanding of the European industrialization process. This is not a survey of comparative studies of technology, nor am I concerned with comparative technology studies in themselves, although there are many interesting and unexplored possibilities. I am more concerned with how the modern history of technology can contribute to wider debates about the pattern of industrialization in Europe.

1. Typologies of European Industrialization

Why should we do comparative studies? Comparative history rests on the idea that regional or national studies can add to or modify a more general theory which is based on a single case, or limited number of cases. Applied to European history of the last century, this means that we would want to use local, regional or national case studies to explore or modify some general approach to how and why the process of modernization and industrialization took place. This immediately raises the questions of what this general approach or understanding actually is. In fact, of course, there is little consensus at the present time even on the characterization of industrialization for single countries, let alone Europe as a whole. A few years ago Patrick O'Brien addressed this issue at length in his article *"Do we have a typology for the study of European industrialization in the 19th century?"*; O'Brien analysed three contenders for the role of "paradigms" of the European industrialization.[o] Others, notably Sidney Pollard, have likewise dealt with this issue, for example in his short book, *Typologies of European Industrialization*.[1] and, of course, the issue is central in current debates on German development, perhaps particularly so in recent research relating to the nature of the German "Sonderweg".[2]

Let me briefly turn to the major paradigms of European industrialization. These are, firstly, the idea that European industrialization involved emulation or imitation of the British model. This is a diffusion approach, where "diffusion" refers to much more than the diffusion of technology – it refers to the adoption of a complex set of economic activities and structures. Secondly, there is the "discontinuity" paradigm, the idea that industrialization involves a fairly sharp "take-off" phase, which has more or less common characteristics across countries. Thirdly, there is an approach based not on nations but on regions as units of analysis. The idea is that the problem is to understand conditions of regional growth; this approach is closely linked with the concept of proto-industrialization.

All of these general approaches have weaknesses which are increasingly recognized. I discuss them briefly, in reverse order:

The paradigm based on European regional development often neglects the impacts of national effects, particularly on trade. The contribution of proto-industrial and other regional rather than national studies to our understanding of nineteenth century developments has long been acknowledged, demonstrated *inter alia* by Sidney Pollard, Pat Hudson, Maxine Berg and others. Industry did not emerge evenly distributed across Europe, but in clusters. Looking to regions we find the social, industrial and economic developments we need to study in order to understand the process of industrialization; developments which may not be visible, or may be conflated from a national point of view. The study of local factors explaining why industry established where it did, grew in some areas and declined in others, is obviously important. But as a potential main pattern, or typology for European nineteenth century industrialization, O'Brian points to a number of serious weaknesses. On the one hand, there are problems of evidence at regional level, particularly statistical evidence: problems of input and output data, of reaggregating data, and generally of providing data sufficient for mapping production trends. Beyond this, the regional approach is one in which trade factor mobility are usually absent. However, nineteenth century European development was characterized by extensive trade between nations; we know that international trade in machinery was extensive and still need, I think, to consider the effects of custom duties policies on the rate and direction of technological change and industrialization in Europe. Moreover there was extensive factor mobility. The patterns of international trade and factor movements suggest perhaps that while a regional approach is an essential element of industrialization studies, it is not an alternative to national studies.

The second paradigm follows the agenda established by Gerschenkron and

Rostow: European economic growth is seen as characterized by "take offs" or "discontinuities" at various times across Europe. Industrialization is a matter of imitating the British model, partly by way of substituting necessary factors where these are unavailable in the original form. Moreover, development is characterized by late-comers reaping the benefits of "catching up", where late-comers are capable of constructing an appropriate institutional framework. A very influential approach, but a model which is very difficult to confirm in the available data: it is extremely problematic to identify and follow phases or trend breaks given the lack of data for industrial production in the last century. Where this has been attempted, or addressed by other means, researchers have usually abandoned the idea of "take offs". O'Brien refers in his article to research on the Habsburg empire, which show long and steady industrial development; we have Austria, Hungary and Scandinavia which went through more than one growth period etc. As to distinct national patterns of industrialization, again, this is difficult to test, especially because we lack data to compare Gerschenkron's prediction of a tendency to large-scale industry, and emphasis on capital goods production in late-comers. Furthermore, there are indications that late-comers did not start by developing capital goods industries, and capital-intensive techniques generally, but emphasized consumer goods, suggesting confirmation of Hoffmann's approach rather than Gerschenkron's.[3]

The weakness of this approach has further been shown by interesting recent study on German development, which raises serious doubts as to the German model of economic growth and by implication calls for a revision of the explanation of Germany's political development, or "Sonderweg".[4] It questions, for example, the validity of Gerschenkron's claim that late-comers depended on the banks and the state for investable capital and entrepreneurship in their industrialization. There was a variety of state attitudes; in the development of electricity supply in southern Germany, for example, the state in some cases gave active support, was uncertain or gave no support in others.[5] At the best it seems now that the role of the state in German technological transformation was very complex and ambiguous. As for the role of the banks in German industrial development it appears necessary to revise the view, stemming from Gerschenkron, that late-comers depended far more on banks for investment and initiatives than earlier developers (such as Britain and Belgium). Self financing was prevalent, and industrial structural change was predominantly a response to technological innovation and market forces. Furthermore, the medium-sized firm remained the most prevalent until 1914.[6] This in turn may well end up calling into question the relevance of the enormous literature on banking in early European industrialization.

The third paradigm is technology diffusion. Here, continental developments were related to the prior industrial revolution in Great Britain. British leadership in the late eighteenth and early nineteenth century is – sometimes implicitly – seen as a consequence of technological change, and consequently European economic history is understood in terms of diffusion of British production techniques and industrial and commercial forms of organization. The idea of a British breakthrough has existed since the work of Toynbee and Mantoux; its development into a model of continental emulation via technology diffusion was most thoroughly proposed by David Landes. A major problem is that few today accept the underlying version of the Industrial Revolution in England. The English model has been sharply revised, particularly by Nick Crafts who has cast major doubts on whether the pattern of late eighteen and early nineteenth century economic growth actually occurred in ways commensurate with significant discontinuities.[7] Further quantitative studies on European economies have shown that England was not the paradigm, but perhaps an extension in European development exhibiting specific, not general, development characteristics. England becomes a special case, not a paradigm. Furthermore, the diffusion process is now seen as much more complex, not relying on a concept of technology as universally applicable or transferable in any simple sense; I will return to this point below.

What these debates mainly demonstrate is the precarious role of established general conceptions of European economic development; some have been killed off, and most shown to contain serious weaknesses. Against this background, to embark on a major project of comparative studies of European technological developments appears fraught with problems. At the present time there is no norm, we do not know what European economic history to relate a history of European technological development to, were we to embark on it.

But perhaps we can side-step this problem; find a way of focussing comparative technological studies which does not demand a relationship with an absent theory. The debates referred to above encompass a much wider field than technological change, involving fairly macro-level economic developments of Europe; indeed one major problem in all of them is that they hardly mention the specifics of the technological change process at all. However, the problem of understanding European industrialization remains, and the debates have a direct bearing on studies of technological developments in three ways – leaving aside for the moment what we do with the question of technological change and economic growth. First, the debates highlight some problems accruing from dealing with nations as units of study in this period; secondly, they make it necessary to examine critically which institutions we focus on, and thirdly, they make it necessary for us to consider more widely the role of technological change, technology diffusion and technology adaptation within European economic history.

2. The role of Technology in European Industrialization

Let me make a link between these very wide problems in the history of European industrialization and the history of technology, by making some brief comments about the role of technological change in industrialization. Something which remains, in my view, worth emphasizing is that economic histories of industrialization more or less completely lack any adequate account of the technological change process. Many early histories of British industrialization were explicitly technological determinist, treating industrialization as an effect of dramatic technological breakthroughs, but never explaining how these technologies emerged. Postwar histories of the European economy tend to get around the problem of technological change simply by mentioning it. It is only very recently that we have had studies, by Inkster and Mokyr, for example, which relate technological change to growth in any consistent and serious way.[8]

Now this really is a glaring weakness, because it is plain that one of the key features of industrialization is the development and diffusion of new technologies. Of course we need to escape from the idea that everything can be explained by such big breakthroughs as steam power, and we must acknowledge that the technological change process was often very slow. But it is an inescapable element in European industrialization and growth. I will say more below about specific research problems which I think are outstanding, but I do want to emphasize that economic histories of European industrialization without a much more complete understanding of the technological base are really like Hamlet without the Prince.

3. Comparative Studies of European History of Technology

Let me turn, at last, to the contribution which I think comparative studies in the history of technology can make to these wider technological issues in the history of industrialization. I want to begin by saying something about the status of the history of technology within historical studies generally; this leads to some suggestions about why comparative studies should be a more explicit part of our research profile.

The accelerating growth in studies in the histories of technology, over the past two decades, has really been much more prominent in the US than in Europe. In Europe the sub-discipline of history of technology has yet to establish itself as a significant component within historical studies. Unlike the US, this is not a field where there is a full-scale professional society, a major journal, widespread teaching, postgraduate research programmes, and so on. What we have is a much more sporadic set of activities, often very exciting in terms of their approaches and quality, but also isolated; historians of technology have not yet succeeded in creating a coherent presence as a discipline.

This is a pity for several reasons. These have to do partly with the significance of technological change in economic development. It seems clear that technological change is a key factor in economic evolution, and as I have emphasized, we still lack an adequate historical account of the relationship between technological change and early industrialization in Europe, or of the dynamics of technological change in later European economic growth. But more important than this, I think, is the conceptual approach taken by the modern history of technology, which is relevant to a wide range of issues within modern history. What I want to suggest here is that one line of development for the history of technology is to take the broad conceptual framework which the subject has developed, and to use this in comparative studies. I think that there is potential here for a rewarding line of research which could give a distinctive focus to European studies in the history of technology.

The basic insight of the modern history of technology is that technology must be understood as a social phenomenon, in the sense that social factors play a powerful role in shaping the evolution of technologies. It is worth saying a little more about the underlying concepts here. Technology consists in part of a body of accumulated knowledge, both formal and informal, about processes of material production. In part, this knowledge relates to the most visible dimension of technology, namely hardware: tools, machinery and equipment, and so on. We can think of a technique as a combination of equipment and skills and rules required for operation. But techniques can only operate within a social framework, which should be understood at different levels. On the one hand there are processes of enterprise and production management, which essentially involve organizations that integrate techniques with each other, and with non-technical activities such as relations with customers. But this takes place within a much wider social context, which shapes and is shaped by the technological change process: a social context of cultural attitudes, social values, political systems and choices, gender roles, income distribution, regional identities, and so on. This wider social context in effect makes choices about such matters as education and literacy, acceptable patterns of working life, and the role of entrepreneurship. These often implicit social choices in turn exert powerful impacts on the process of technological change: on its scope, its characteristics, and on the rate at which change can occur. The technological choices in turn affect economic outcomes.

Now an obvious point about European society is the extreme variety and diversity of its socio-cultural contexts, both in national and regional terms. These are very considerable variations in all of the social factors sketched above, complicated by major linguistic differences. It is worth noting that this is not a matter of geographically well-defined socio-cultural blocks; quite often, very different cultures coexist in very close proximity within narrow national or regional boundaries; Belgium and the Netherlands are examples. We all know that within apparently uniform national societies there exist, in fact, a range of more or less sharp cultural differences.

What problems are involved in the relationship between this cultural diversity and the process of technological change sine the late eighteenth century? One way of looking at this is to sketch some of the main technological questions within a history of European industrialization; I do not need to emphasize that what follows is very tentative.

It seems to me that any serious account of European economic growth must consider three broad technological issues:

- first, the years after 1845 saw very substantial growth in British capital goods exports to many parts of Europe. This is a technology diffusion process on a large scale; we understand very little of the nature of this trade, about the pattern and content of exports, about what made countries or regions receptive to British exports. We understand little of the nature or the complexity of this diffusion process. It seems obvious that there were, in Europe, both wide differences and unexpected similarities in the ability to establish specific industries and to use specific foreign technologies; these patterns presumably owe something to specific features of the socio-cultural environment. How can this be seriously studied, if not in a comparative context?
- second, one component of the industrialization was the indigenous production or international diffusion of essentially similar process technologies. Technologies such as steam power, powered textile equipment, and metal-working techniques in the early phase of industrialization; electrical generation and electrical machinery, or internal combustion engines, in the late nineteenth and early twentieth centuries; computers and modern telecommunications technology at the present were closely similar. These are, in effect, "standard" process technologies; they are technically very similar if not the same. How did very different national and regional entities adapt to and use these technologies? The problem is, what happens when a "standard" technology meets a differentiated culture? What is it about apparently very different social and cultural contexts which makes such technologies feasible; how do societies and cultures adapt and change where that is demanded by the characteristics of the technology?
- third, particularly at regional level, highly differentiated patterns of technical

capability and practice emerged or were sustained in the 19th century. Piore and Sabel's concept of "flexible specialization" actually points to the pattern of very specific technical competences, and we know very little about how these emerged, and how they were integrated with the wider national and international economy. One of the weaknesses of the Piore and Sabel's work is that it was based essentially on rather *ad hoc* examples – it would be going too far to call them case studies. It is often difficult to assess how generally applicable their examples really are. Here, comparative studies are potentially very fruitful, if we want a more complete understanding of the significance of these highly localized patterns of technology use within the context of apparently very different local societies and cultures.

The point I am making, of course, is that these major questions require a comparative perspective if they are to be answered. They require collaboration or integration between quite detailed studies within different European environments. Systematically looking at such issues would be a research programme of great intrinsic interest, but we should note that such research would also have two further implications for the history of technology in Europe. First, because the degree of socio-cultural diversity in Europe is almost unique in the world, comparative studies which set technological change within this context would give European history of technology a unique and very interesting research profile; what I am suggesting, therefore, is continuing to relate closely to the achievements of our fellow historians of technology in the USA, but going in quite a different research direction. Secondly, the comparative approach would establish the history of technology more firmly within historical studies as a whole. It is not too much to hope that the significance of answers to the questions posed above would go well beyond the history of technology. I have suggested that the history of European industrialization is in something of an impasse, both conceptually and empirically. Comparative work in the history of technology is not by itself going to solve any of the big problems, but it might well constitute an important step towards clarifying the issues, and should certainly be a central component in any future historical synthesis.

Notes

0. P. K. O'Brien, "Do we have a typology for the study of European industrialization in the XIX-the century?", Journal of European Economic History, Vol 15 No 2, 1986, pp.291-333.
1. Sidney Pollard, *"Typology of Industrialization Processes in the Nineteenth Century"* (No.39, Fundamentals of Pure and Applied Economics), Harwood Academic Publishers, 1990.
2. W. R. Lee (ed.), "German Industry and German Industrialisation", Routledge, London 1991
3. see O'Brian's discussion, op. cit. pp. 310-112.
4. W. R. Lee, *"German Industry and German Industrialisation. Essays in German economic and business history in the nineteenth and twentieth centuries"*, Routledge, London 1991.
5. See Herman Schaefer's chapter in W. R. Lee (ed.) op.cit.
6. See Wilfried Feldenkirchen's chapter in W. R. Lee (ed.), op.cit.
7. Nick Crafts, *"British Economic Growth during the Industrial Revolution"*, Oxford University Press, 1985
8. Joel Mokyr, *"The Lever of Riches"*, Oxford 1990; Ian Inkster, *"Science and Technology in History"*, Macmillan, 1991. See also P. Mathias and J. A. Davis (eds.), *"Innovation and Technology in Europe from the Eighteenth Century to the Present Day"*, Blackwell, 1991

Contributors

Martyn Bakker
Born 1956. He studied medieval history at Nymegen University. In 1983 he joined the History of Technology Section of the Faculty of Philosophy and Social Sciences at Eindhoven University of Technology, where he wrote a thesis on Dutch beet sugar industry in the 19th and early 20th centuries. From its start in 1988 he was involved in a very large, Eindhoven-based research project on technology and industrialisation in the Netherlands. The results of this project will be published in a six volume history of technology, *'Geschiedenis van de Techniek in Nederland. De wording van een moderne samenleving, 1800-1890'* (Zutphen, De Walburg Pers). The first volume, on agriculture and food, containing articles on dairying, sugar industry, milling of grain, and introductory and concluding chapters by Martyn Bakker, appeared in November 1992.

Kristine Bruland
Born 1950 in Oslo. Educated mainly in England, and holds a D.Phil. in Modern History from the University of Oxford. Her research interests lie within the fields of social and economic history and the history of technology. Publications include 'Industrial conflict as a source of technical innovation: three cases', *Economy and Society*, Vol. 3 No. 2, 1982; *British Technology and European Industrialization. The Norwegian textile industry in mid-nineteenth century*, 1989, (Cambridge: Cambridge University Press); and (ed.) *Technology Transfer and Scandinavian Industrialisation*, 1991 (Oxford, New York: Berg Publishers). She is presently working in the Department of History, University of Oslo.

Dan Ch. Christensen
Born 1941 in Copenhagen. Graduated from the University of Copenhagen 1970. Lecturer at the University of Roskilde from 1972. Assistant professor since 1976. His work has concentrated on teacher education (history) and internationalisation of university courses. His research has been in the history of ideas, history of agriculture and history of science and technology. From 1990 head of the TISC group (Technology, Innovation, and Society in a Cultural perspective, set up by

the National Research Council) writing a 3-volume Danish History of Technology 1750-1990. Has published on the history of education, history of ideas, history of science and technology (particularly Grundtvig, Oersted, steam engines, agriculture, canals, metallurgy).

Henrik Harnow

Born 1961 in Odense. Part-time lecturer in history at Odense University 1989-92. He has mainly been teaching English history, concentrating on the technological aspects of the Industrial Revolution as well as social and economic history. His research has been in the field of industrial archaeology, especially the evolution of the factory system in Denmark during the 19th century and the transfer of textile technology to Denmark. At present he is a research fellow of the TISC-project. He is primarily dealing with Danish engineering history from the late 19th century to the present, in particular engineering education and the complex process of professionalization.

John R. Harris

Graduated from Manchester University and wrote his Ph.D. thesis there. Reader in Economics and Director of Social Studies at Liverpool University,1953-70. Head of Department of Economic and Social History at Birmingham University from 1970. Emeritus Professor 1990. One of the founders of the Ironbridge Institute 1980-90. Received the French Order 'Chevalier des arts et des lettres' 1990. Major works: *'A Merseyside Town in the Industrial Revolution'* (with prof. T.C. Barker), 1954, *'Blue Funny'*, 1956, *'The Copper King'*, 1964, *'Liverpool and Merseyside'*, (ed.), 1969, *'Essays in Industry and Technology in the Eighteenth Century'*, 1992.

Hans Hedal

Born 1955 in Copenhagen. Graduated from Roskilde University Centre (history and physics) 1987. Research librarian 1987-88. Gymnasium teacher 1989-90. Assistant lecturer at Roskilde University Centre 1987-89 and 1990-91. From 1991 research fellow studying Danish history of energy technology at the TISC- project (Technology, Innovation and Society in a Cultural perspective, sponsored by the National Research Council). This is aiming at writing a 3-volume history of Danish Technology and Culture. He is presently working on a major thesis about the unsuccessful Danish nuclear power project.

Peter Kemp
Born 1937. Professor of Philosophy at the University of Copenhagen since 1972. Visiting Professor at the University of Gothenburg, Sweden 1987-88, and at the Technical University of Vienna winter 1988-89 and summer 1990. President for the *Nordic Institute of Philosophy* 1980-89. Member of *The Internal Academy of Philosophy of Science* since 1984 and *The International Institute of Philosophy* since 1989. Member of different commissions for the Danish Government: *On Technological Risks* (1988) and on *Research involving Human Subjects* (1989).

Major Works: *Théorie de l'engagement, I-II*, Seuil, Paris; *Éthique et Médecine*, Tierce et INSERM, Paris 1987; co-editor of *Technologies et Sociétés*, Galilée, Paris, 1980; *The Narrative Path. The later Works of Paul Ricœur*, MIT-Press, Mass., 1989.

Helge Kragh
Born 1944 in Copenhagen. Senior research fellow at the TISC-project, Roskilde University Centre. Research and publications in the history of modern physical sciences, electrical communication technology and the science-technology relationship. Books include *An Introduction to the Historiography of Science* (1987), *Dirac: A Scientific Biography* (1991), both published by Cambridge University Press, and a forthcoming work on the historical development of long-distance telephony in Europe.

Dick van Lente
Born 1952 in Djakarta, Indonesia. He studied modern and theoretical history at the University of Amsterdam. He teaches cultural history of industrial societies at the Erasmus University in Rotterdam, The Netherlands, since 1980. His dissertation, published in 1988, concerns reactions to technological innovations between 1850 and 1920 in the Netherlands. Currently he is extending this research to the period 1750-1850 and trying to put it in a comparative perspective. He is also working on the role of communication technologies in cultural changes during the nineteenth century. Is is one of the editors of a series of books about technological innovations in The Netherlands during the nineteenth century, the first part of which will appear in the fall of 1992.

Klaus Mauersberger
Born 1950 in Annaberg-Buchholz. Graduated engineer (marine engineering) 1972, dissertation in applied mechanics 1979, teaching assistant in the history of technology at Dresden University of Technology since 1977. His work has focused on the history of the engineering sciences, especially the history of applied mechanics

and mechanical engineering. His research has been in the relationship between technology and culture, the history of the education of engineers in a European context, the epistemological roots of the engineering sciences and the history of mechanics and machines and includes contributions to several books (e.g. *Lebensbilder von Ingenieurwissenschaftler*, 1989; *Geschichte der Technikwissenschaften*, 1990) as well as a number of articles related to the main topics mentioned.

Akos Paulinyi
Born 1929 in Budapest. Professor of the history of technology and of economic history at the Technische Hochschule Darmstadt in Germany since 1977. Up to the 1970s his publications covered the field of the economic history of Slovakia and of the Austro-Hungarian Monarchy in the 18th and 19th century. In the last decade the bulk of publications is in the field of the history of technology (e.g. technology during the Industrial Revolution in Britain, iron-making, metal-working and the transfer of technology).

Frank Allan Rasmussen
Born 1951 in Copenhagen. From 1989 assistant curator at The Royal Naval Museum in Copenhagen. His research has been in the history of the military industrial complex and its influence on technological development in Denmark. He has published articles on naval architecture, steam engines and industrial archaeology. At present he is a research fellow at the TISC-project, personally working in the field of maritime technology.

Geert Verbong
Born 1955 at Tegelen, The Netherlands. Lecturer in the history of technology at Eindhoven University of Technology since 1990. After studies in physical engineering, he specialized in the history of technology. Since 1982 he is involved in a research project on the history of technology in The Netherlands in the 19th century. For this project he worked on the cotton printing and dyeing industry, concluded with a Ph. D. thesis on this subject. From 1988 until 1993, he is doing research on the science-technology relationship and the rise of the technical profession in the Netherlands. The whole project will come to a close in 1994 after the publication of the last of six volumes on 'Technology in The Netherlands, the becoming of a modern society 1800-1890'.

Michael F. Wagner

Born 1956. Research Fellow, editor of the periodical, *Den jyske historiker.* 1987-1991 assistant professor at the Technical University of Aalborg. Since 1991 at the Department of History, University of Aarhus. At present working on a major thesis on polytechnical education in Denmark during the 19th century. Has published several articles on history and technology during the past ten years.

Index

agricultural technology, 10f., 15, 102ff., 107f., 175ff., 189ff.
Aristotle, 218
artisans, 19ff., 38, 42, 45, 156ff., 196, 215f.
Ashton, T.S., 197
atomic bomb, 210
Austria, 118, 228

Bakker, Martyn, 15, 175ff., 235
Beckmann, Johann, 148
Belgium, 19, 119, 138, 228
Beuth, P.C.W., 24, 28
bolts-and-nuts history of technology, 13
British history of technology, 12
British technology, 17ff., 30ff., 42, 47ff., 193f.
Bruland, Kristine, 15, 226ff., 235
butter, 180ff.

Cameron, R., 19
Canguilhem, Georges, 221
Christensen, Dan Ch., 9ff., 189ff., 235
Christianity, 218f.
civil engineering, (*see* engineering history)
Club of Rome, 210
colonies (Danish-Norwegian), 42
comparative history of technology, 12, 15, 198, 226ff., 230ff.
computerization, 212f., 232
cooperative societies, 181ff., 189f., 199ff.
Crafts, Nick, 229
craftsmen, *see* artisans
cream separator, 180, 200
cultural construction of technology, 10ff., 84ff., 160, 176ff., 187f., 194, 198, 230ff.

Danish history of technology, 12, 16, 93
Danish Royal Naval Dockyard, 13, 41ff.
Denmark, 73, 78, 41ff., 91ff., 146ff., 164ff., 180, 188
diffusion of technology, 18, 229

Dutch history of technology, 12, 16

education, (*see* engineering history, polytechnical education, technical education)
economic historians, 9f., 175, 187f., 193f., 226ff.
electrification, 14, 91ff., 232
electrophones, 80
Ellul, Jacques, 213
Engels, A.H.J., 63f.
engineering history, 14, 111ff., 125ff., 146ff., 164ff., 197
Enlightenment, the, 112
ethics, 15, 204ff.
Europe, 70ff., 226ff.

Fischer, Claude, 87
France, 13, 19, 30ff., 46f., 75, 111ff., 119, 128ff.,152, 185
French history of technology, 13
Freud, S., 206
German history of technology, 12, 16, 17ff.
Germany, 17ff., 58, 75, 111, 116ff., 138, 146, 148, 154, 167, 168, 172, 185, 228
Gerner, Henrik, 48
Gerschenkron, A., 227
Grundtvig, N.F.S. 189, 198, 201

Hagemann, G.A., 170
Hannover, H.I.,169f.
Harnow, Henrik, 14, 164ff., 236
Harris, John R., 13, 30ff., 46, 52, 236
Hedal, Hans, 14, 91ff., 236
Hegel, G.W.F., 206f.
Heidegger, M., 213f., 216f., 223f.
historical source material, 30ff.
historiography, 9ff., 226ff.
Hodgkinson, Christopher, 219
Hohlenberg, F.C.H., 48
Holland, *see* Netherlands, the

Index

Holmberg, L.F., 165, 167
Hungary, 228
Huxley, Aldous, 212

ideology, 56f.
industrial espionage, *see* technological espionage
Inkster, Ian, 230
innovation, 18, 41ff., 52, 56, 194, 197, 200
invention, 68, 160
Italy, 47

Jeremy, D.J., 23

Kant, I., 206, 219f.
Kemp, Peter, 15, 204ff., 237
'kindergarten method', 20
Krabbe, F.M., 47f.
Kragh, Helge S., 13f., 68ff., 237
Kuyper, Abraham, 61f.

Lacan, J., 206
la Cour, Paul, 14, 68, 91ff.
Landes, David, 229
learning-by-doing, 20
Lente, Dick van, 13, 16, 55ff., 237
Lindqvist, Svante, 52
Lundgreen, Peter, 111, 125

'machine age', 57
machine tools, 17ff., 50f.
maritime technology, 41ff., 51ff.
Marx, Karl, 209
Mauersberger, Klaus, 14, 111ff., 237f.
mercantilism, 10
Mirandola, Pico della, 220
military engineering, 41ff., 130ff.
Mitchell, Andrew, 49
Mokyr, Joel, 230
modernization process, 9ff., 41ff., 52, 189ff., 202
Myllyntaus, Timo, 18, 74, 87

Netherlands, the, 42, 47, 55ff., 119, 125ff., 175ff.
Nielsen, L.C., 200
non-scientific origin of technology, 20, 215f.

Norwegian history of technology, 12, 16
nuclear power, 91ff., 208f., 211f., 222
O'Brien, Patrick K., 226f.
Oersted, H.C. (*see* Ørsted, H.C.)
Orwell, George, 213

Parker, H.T. 32
Paulinyi, Akos, 13, 17ff., 238
Petersen, P.O., 170
ploughs, 11f., 189f., 200
Pollard, Sidney, 22, 28, 226
polytechnical education, 14, 119, 125ff., 134f., 138ff., 146ff., 195f., 198
professionalization, 45, 159, 165ff.
proto-industrialization, 227
Prussia, 17ff.

quality of life, 204ff.

Radkau, Joachim, 12
Randall, Adrian, 56
Rasmussen, Frank Allan, 13, 42ff., 238
Redlich, Hans, 20
Robinson, Eric, 35
Romantic-Conservative criticism of technology, 14, 55ff.
Rolt, L.T.C., 31
Roskilde conference, 12ff.
Rostow, W.W., 228
Russia, 42, 48, 119

Saxonia, 19, 171
Sartre, J.P., 223f.
Scandinavia, 228
science and technology, 45, 112ff., 125ff., 141, 147, 151f., 192, 197, 201, 215
science and values, 184ff.
social construction of technology (*see also* cultural construction of technology), 10
Sorge, Frederik van, 55, 60f.
Spengler, Oswald, 209
steam engines, 49, 180f., 201, 232
sugar, 176ff.
Sweden, 42, 48, 74f., 165
Swedish history of technology, 12, 16, 52
Switzerland, 19, 58, 74f., 118, 119, 168

technical education (*see also* engineering history *and* polytechnical education), 44, 111ff., 125ff., 146ff., 164ff.
technical journals, 20f., 150f.
technological espionage, 13, 22f., 30ff., 41, 46ff.
technological optimism, 209
technology, concept of, 215f.
technology transfer, 9ff., 17ff., 26ff., 41ff., 68ff., 85f., 192f., 198f.
telegraph technology, 76f.
telephone technology, 14, 68ff., 76f.
test-tube-babies, 212
theatrophones, 80
theory and practice tension, 125ff., 139, 142f., 157ff.
Tipton, F.B., jr., 28
TISC-Project, 12, 16
TMV-Centre, Oslo, 12, 16

tools act, 22f., 34f., 49f.

UK, 11, 20ff., 30ff., 47ff., 58, 72f., 78, 115f., 119, 120, 165, 169f., 189, 228
Ursin, G.F., 150ff., 197
USA, 68ff., 118, 168, 172, 230

Verbong, Gert, 14, 125ff., 238
Vlekke, J.F., 186f.

Wagner, Michael F., 14, 146ff., 239
wind-electricity, 91ff., 101ff.
Winstrup, Ole, 154, 160, 197
Winstrup, P.J., 200

Young, Alexander, 49

Ørsted, H.C., 149ff., 164f., 189, 195ff.